Engineering Optimization: Methods and Applications

Optimization carries great significance in both human affairs and the laws of nature. It refers to a positive and intrinsically human concept of minimization or maximization to achieve the best or most favorable outcome from a given situation. Besides, as the resources are becoming scarce there is a need to develop methods and techniques which will make the systems extract maximum from minimum use of these resources, i.e. maximum utilization of available resources with minimum investment or cost of any kind. The resources could be any, such as land, materials, machines, personnel, skills, time, etc. The disciplines such as mechanical, civil, electrical, chemical, computer engineering as well as the interdisciplinary streams such as automobile, structural, biomedical, industrial, environmental engineering, etc. involve in applying scientific approaches and techniques in designing and developing efficient systems to get the optimum and desired output. The multifaceted processes involved are designing, manufacturing, operations, inspection and testing, forecasting, scheduling, costing, networking, reliability enhancement, etc. There are several deterministic and approximation-based optimization methods that have been developed by the researchers, such as branch-and-bound techniques, simplex methods, approximation and Artificial Intelligence-based methods such as evolutionary methods, Swarm-based methods, physics-based methods, socio-inspired methods, etc. The associated examples are Genetic Algorithms, Differential Evolution, Ant Colony Optimization, Particle Swarm Optimization, Artificial Bee Colony, Grey Wolf Optimizer, Political Optimizer, Cohort Intelligence, League Championship Algorithm, etc. These techniques have certain advantages and limitations and their performance significantly varies when dealing with a certain class of problems including continuous, discrete, and combinatorial domains, hard and soft constrained problems, problems with static and dynamic in nature, optimal control, and different types of linear and nonlinear problems, etc. There are several problem-specific heuristic methods are also existing in the literature.

This series aims to provide a platform for a broad discussion on the development of novel optimization methods, modifications over the existing methods including hybridization of the existing methods as well as applying existing optimization methods for solving a variety of problems from engineering streams. This series publishes authored and edited books, monographs, and textbooks. The series will serve as an authoritative source for a broad audience of individuals involved in research and product development and will be of value to researchers and advanced undergraduate and graduate students in engineering optimization methods and associated applications.

Chetan B. Khadse · Ishaan R. Kale ·
Apoorva S. Shastri
Editors

Intelligent Methods in Electrical Power Systems

Springer

Editors
Chetan B. Khadse
Department of Electrical and Electronics Engineering
Dr. Vishwanath Karad MIT World Peace University
Pune, Maharashtra, India

Ishaan R. Kale
Institute of Artificial Intelligence
Dr. Vishwanath Karad MIT World Peace University
Pune, Maharashtra, India

Apoorva S. Shastri
Institute of Artificial Intelligence
Dr. Vishwanath Karad MIT World Peace University
Pune, Maharashtra, India

ISSN 2731-4049 ISSN 2731-4057 (electronic)
Engineering Optimization: Methods and Applications
ISBN 978-981-97-5717-6 ISBN 978-981-97-5718-3 (eBook)
https://doi.org/10.1007/978-981-97-5718-3

This Springer imprint is published by the registered company Springer Nature Singapore Pte Ltd.
The registered company address is: 152 Beach Road, #21-01/04 Gateway East, Singapore 189721, Singapore

Preface

In today's era intelligent methods have become a part and parcel of electrical power systems. One cannot think of power systems without the smart technologies. The microgrids and smart grids are found to be very popular and practical with the help of renewable energy integration. The intelligent methods are widely used in generation, transmission, and distribution systems. The intelligent generation systems desire to produce fuel flexible, cleaner, and adaptive electrical power. In case of transmission, intelligent systems can be implemented for the smooth transmission of electricity from generation to consumer end. It can be used in power quality analysis, fault analysis, load flow studies, etc. The planning, operation and control are the major areas of electrical distribution systems in which intelligent systems can be incorporated. Along with these home automation, autonomous electric vehicles, smart charging stations, energy theft detection, renewable energy integration, and industrial automation are the points of interest in which artificial intelligence can be applied. It can be said that intelligent methods are the boon for the electrical field which may enhance the quality of the power sector and contribute to the social benefit.

This edited book aims to discuss and exhibit the concepts of intelligent methods used in electrical power system applications. These intelligent methods include optimization methods, learning algorithms, intelligent home automation methods, gradient descent algorithms, control algorithms, etc. This book also addresses the mathematical modeling of the electrical power system components which is the basic requirement of the application of the intelligent methods to the electrical power systems. The chapters which are submitted to the edited book are critically reviewed by at least two reviewers. The critical and valuable suggestions provided by the reviewers enhanced the quality of the chapter in terms of methodology, performance of the critical discussion, depiction and quality of the solution. These suggestions surely contributed to the future artificial intelligence trends in the electrical power system domain. This edited book is a valuable asset for the researchers, professors, and industry professionals working in the field of electrical engineering.

In Chap. 1 "A Comprehensive Review on Metaheuristic Optimization Methods for Efficient Power System Operation", Ankur Maheshwari et al. confer about a comprehensive overview of metaheuristic methods for addressing optimization issues in power systems. It explores the most prevalent metaheuristic optimization approaches and their practical application in solving power system optimization problems. Additionally, it highlights common misconceptions found in the existing literature and provides practical suggestions for developing effective strategies to apply metaheuristic algorithms in tackling power system challenges. By shedding light on the capabilities and potential of metaheuristic algorithms, this article aims to foster further advancements in the field of power system optimization.

In Chap. 2 "An Application of Artificial Bee Colony and Cohort Intelligence in the Automatic Generation Control of Thermal Power Systems" Murugesan Dhasagounder et al. discuss about the artificial bee colony and cohort intelligence optimization techniques for the application in thermal power system. A comparison between the performance of artificial bee colony (ABC) optimized controllers and cohort intelligence optimization (CIO) optimized PID controllers in the context of the interconnected reheat thermal power system is made. The findings of the study indicate that PID controllers optimized using the CIO method outperform those optimized using the ABC method in the context of the mutually dependent reheat thermal power system. The aim of this research work is to fine-tune the parameters of a Proportional-Integral-derivative (PID) controller for automatic generation control in a mutually dependent two-area power system with reheat thermal power generation. Both ABC and CIO are employed to determine the optimal gain values for the controller, involving the evaluation of multiple cost functions.

In Chap. 3 "Integration of Intelligent Systems for Efficient Smart Grid Management". O. Apata and Pitshou N. Bokoro elaborate about the role played by intelligent systems, especially multi-agent systems in the provision of extensible and flexible architecture for the deployment of intelligence in a smart grid. The authors mentioned about the implementation of advanced technologies and communication systems that provide greater control, efficiency, reliability, and sustainability in the generation, transmission, distribution, and consumption of electricity. The integration of intelligent systems plays a crucial role in the effective management of the smart grid.

In Chap. 4 "Investigation of Electric Load Forecasting Methods: A Weka Application (Regression and Optimization)" Düzgün Akmaz proposed an application for the forecasting of an electric load which is simple and effective method. The highest success with the least input features was aimed in the methodology. The regression feature of the Random Forest (RF) algorithm and the Correlation Attribute Eval (CAE) feature selection methods were used to achieve this. The simulation results proved that the proposed method was successful. While WEKA program was used for RF and CAE algorithms, real-time data was obtained from Spain.

In Chap. 5 "Scaled Conjugate Gradient-Based Intelligent Microgrid Fault Analysis System", Nishant Chaudhari et al. proposed an artificial neural network-based fault analysis system for the transmission line of the power system. In this chapter, artificial neural network is trained with the scaled conjugate backpropagation algorithm with the input fault data. The dataset for the training is generated through MATLAB Simulink. A simulation model is built with the help of numerical parameters and voltage, currents are acquired through measurement blocks. This data is used for the training using NPRTOOL blockset. The detected and classified results are presented in the chapter.

In Chap. 6 "IoT-Based Intelligent Home Automation System Using IFTTT with Google Assistant", Mohan Bansal et al. discuss about an IoT-based IHAS prototype designed using IFTTT (If This Then That) and Google Assistant. The system allows users to control and automate various devices in their home through voice commands in multiple languages given to Google Assistant or through Adafruit IO from anywhere on the earth. The system integrates various alternating current (AC) and direct current (DC) devices such as electric bulb, electric socket, DC motor, and light emitting diodes (LEd) lights. This system allows them to communicate with each other to perform tasks based on user preferences. The system operates on the logic created in IFTTT, where the user can set up conditions or triggers for a specific action to take place. This allows for a personalized and flexible home automation experience for the user. The integration with Google Assistant further enhances the ease of use, allowing the user to control their smart home devices through simple voice commands in any language. The multilingual voice control system may enhance the system's robustness and performance.

In Chap. 7 "Scaled Conjugate Gradient Backpropagation-Based Fault Analysis System for Induction Motor" Chaitanya Nimbargi et al. proposed a fault analysis system for induction motor based on scaled conjugate gradient backpropagation. An induction motor is simulated in MATLAB with the specifications. Faults are created in the simulated induction motor as a part of data generation process for the artificial neural network. The fault data is used for the supervised learning which is the input for the training. As mentioned SCGB neural network is used as a learning algorithm. The data is divided into training, testing and validation data. The training performance is presented in the paper. The faults considered are symmetrical as well as unsymmetrical faults. The detected results for line to ground, line to line, double line, double line to ground, triple line, triple line to ground faults are analyzed and presented in this paper.

In Chap. 8 "Ice Thickness Control Circuit to Automate the Milk Chilling System", Kalash Nikhar et al. proposed a control circuit based on a control algorithm to automate the process of milk chilling system. This chapter presents the development and implementation of an ice thickness controller circuit that starts the cooling as milk can is placed in the chilling system and stops the cooling as soon as milk is cooled to desired temperature and ready for the transportation. An ice thickness sensor is designed using the principle of conductivity. Outcome of the work is an automated milk cooling process that enables energy saving by running compressor on and off on reaching the thresholds of the temperatures ensuring the required ice level.

In Chap. 9 "Real-Time Monitoring and Battery Life Enhancement of Surveillance Drones", Pooja Kumari et al. address these issues by introducing creative approaches and novel ideas to improve surveillance drone architecture. It discusses cutting-edge technology such as edge AI, which involves establishing artificial intelligence directly on drones, innovative drone propeller designs, various AI algorithms, and fresh approaches to drone architecture. These improvements aim to make drones capable of performing difficult and vital jobs in addition to ordinary surveillance. The ultimate goal is to improve the efficiency and practicality of surveillance drones, allowing them to succeed in a variety of applications and circumstances.

Dr. Chetan B. Khadse
Assistant Professor
Department of Electrical and Electronics Engineering
Dr. Vishwanath Karad MIT World Peace University
Pune, Maharashtra, India
chetan.khadse@mitwpu.edu.in

Dr. Ishaan R. Kale
Research Assistant Professor
Institute of Artificial Intelligence
Dr. Vishwanath Karad MIT World Peace University
Pune, Maharashtra, India
ishaan.kale@mitwpu.edu.in

Dr. Apoorva S. Shastri
Research Assistant Professor
Institute of Artificial Intelligence
Dr. Vishwanath Karad MIT World Peace University
Pune, Maharashtra, India
apoorva.shastri@mitwpu.edu.in

Contents

Editors and Contributors

About the Editors

Chetan B. Khadse holds a Ph.D. in "AI applications in Electrical Engineering" from Visvesvaraya National Institute of Technology Nagpur, India, an M.Tech. in Power Systems from Shivaji University, India, and a Bachelors from Amravati University, India. He is currently working as an Assistant Professor in the School of Electrical Engineering at the MITWPU, Pune, India. His research interests include artificial intelligence, power systems, power quality, optimization algorithms, and intelligent systems. He has published over 18 research papers in peer-reviewed reputed journals, chapters, and conferences. He is actively involved in the Center of Excellence of Electric vehicle at MITWPU where his research is mainly focused on AI techniques for the betterment of EVs.

Ishaan R. Kale holds a Ph.D. in Nature-Inspired Optimization Techniques from the faculty of Mechanical Engineering at Symbiosis International University. He received his Master of Engineering in Mechanical Design Engineering from Maharashtra Institute of Technology, Pune University, and his Bachelor of Engineering from North Maharashtra University. Ishaan worked as an Assistant Professor at Symbiosis Institute of Technology for six years. Currently, he is working as a Research Assistant Professor at the Institute of Artificial Intelligence, MIT World Peace University. His research interests include Design Engineering, Structural Optimization, Computational Intelligence, Constraint Handling, Probability Collectives, Socio Inspired Optimization Methods, Physics-Based Optimization Methods, Cohort Intelligence, Particle Swarm Optimization, Genetic Algorithms, Hybrid Metaheuristics, Game Theory, Operation Research, and Numerical Methods. He has published 10 research papers in peer-reviewed journals, conferences, and chapters along with one authored and one edited book. He is also involved as an active research member of the Optimization and Agent Technology Research Lab.

Apoorva S. Shastri holds a Ph.D. in Optimization Algorithms and Applications from Symbiosis International (Deemed University), a Master of Technology (M.Tech.) in VLSI Design, and a Bachelor of Engineering in Electronics and Product Design Technology from R.T.M.N.U, Nagpur. She has also earned a Diploma from the Governemt Polytechnic, Nagpur. She worked as a guest faculty at the Centre for Development of Advanced Computing (C-DAC), Pune. Currently, she is a Research Assistant Professor at the Institute of Artificial Intelligence at the MITWPU, Pune, India. Her research interests include optimization algorithms, VLSI design, multi-objective optimization, continuous, discrete, and combinatorial optimization, complex systems, manufacturing, and self-organizing systems. Apoorva developed socio-inspired optimization methodologies such as Multi-Cohort Intelligence Algorithm and Expectation Algorithm. Apoorva has published several research papers in peer-reviewed journals, chapters, and conferences along with one Springer authored book. She is a regular reviewer of different journals Elsevier and Springer. She has also served as session chair for a few international conferences.

Contributors

Düzgün Akmaz Engineering Faculty, Department of Electrical and Electronics Engineering, Munzur University, Tunceli, Turkey

O. Apata Trilateral Research Chair in Transformative Innovation, The4IR and Sustainable Development, College of Business and Economics, University of Johannesburg, Johannesburg, South Africa

Mohan Bansal Electronics and Communication Engineering, Indian Institute of Information Technology, IIIT Sonepat, Khewra, Haryana, India

Pitshou N. Bokoro Trilateral Research Chair in Transformative Innovation, The4IR and Sustainable Development, College of Business and Economics, University of Johannesburg, Johannesburg, South Africa

Nishant Chaudhari Dr. Vishwanath Karad MIT World Peace University, Pune, Maharashtra, India

Sharvari Dhamale Dr. Vishwanath Karad MIT World Peace University, Pune, Maharashtra, India

Murugesan Dhasagounder Paavai Engineering College, Namakkal, Tamil Nadu, India

Harshith Sourav Gosula School of Electrical Engineering, Dr. Vishwanath Karad MIT World Peace University, Pune, India

Supriya Jaiswal Department of Electrical Engineering, National Institute of Technology, Hamirpur, HP, India

Abhidnya Kadu Eindhoven University of Technology, Eindhoven, Netherlands

Jagatheesan Kaliannan M. Kumarasamy College of Engineering, Karur, India

Piyush Khadke Black & Veatch, Pune, India

Chetan B. Khadse Dr. Vishwanath Karad MIT World Peace University, Pune, Maharashtra, India

Krishna VIT-AP University, Amaravati, Andhra Pradesh, India

Pooja Kumari School of Electrical Engineering, Dr. Vishwanath Karad MIT World Peace University, Pune, India

Netra Lokhande School of Electrical Engineering, Dr. Vishwanath Karad MIT World Peace University, Pune, India

Anand Mahajan Delft University of Technology, Delft, The Netherlands

Ankur Maheshwari Department of Electrical Engineering, National Institute of Technology, Hamirpur, HP, India

Amit Nehete Dr. Vishwanath Karad MIT World Peace University, Pune, India

Kalash Nikhar Dr. Vishwanath Karad MIT World Peace University, Pune, India

Chaitanya Nimbargi Dr. Vishwanath Karad MIT World Peace University, Pune, Maharashtra, India

Ruchir Nimgaonkar Dr. Vishwanath Karad MIT World Peace University, Pune, India

Shivankar Patil Dr. Vishwanath Karad MIT World Peace University, Pune, India

Himanshu Raj Department of Electrical Engineering, National Institute of Technology, Hamirpur, HP, India

Sidharth Sabyasachi Yeungnam University, Gyeongsan, Korea

Ravi Sekhar Symbiosis Institute of Technology (SIT) Pune Campus, Symbiosis International (Deemed University) (SIU), Pune, India

Pritesh Shah Symbiosis Institute of Technology (SIT) Pune Campus, Symbiosis International (Deemed University) (SIU), Pune, India

Rajeev Sharma VIT-AP University, Amaravati, Andhra Pradesh, India

Yog Raj Sood Department of Electrical Engineering, National Institute of Technology, Hamirpur, HP, India

Ayush Yadav VIT-AP University, Amaravati, Andhra Pradesh, India

Chapter 1
A Comprehensive Review on Metaheuristic Optimization Methods for Efficient Power System Operation

Ankur Maheshwari, Supriya Jaiswal, Yog Raj Sood, Himanshu Raj, and Sidharth Sabyasachi

Abstract Optimization is a vital practice to achieve the best possible outcomes by modeling systems to meet management and technical requirements. However, traditional optimization approaches may fall short in scenarios with incomplete or insufficient data. In such cases, metaheuristic algorithms offer a promising solution by providing satisfactory results. These algorithms have gained significant attention in the electricity system domain, as evidenced by the growing number of articles documenting their successful application in academic literature. This article presents a comprehensive overview of metaheuristic methods for addressing optimization issues in power systems. It explores the most prevalent metaheuristic optimization approaches and their practical application in solving power system optimization problems. Additionally, it highlights common misconceptions found in the existing literature and provides practical suggestions for developing effective strategies to apply metaheuristic algorithms in tackling power system challenges. By shedding light on the capabilities and potential of metaheuristic algorithms, this article aims to foster further advancements in the field of power system optimization.

Keywords Metaheuristic algorithms · Heuristic algorithms · Power system · Efficiency

1.1 Introduction

The optimization problem entails the selection of the most suitable option among a set of alternatives, aiming to maximize a specific objective while simultaneously minimizing associated costs or expenses (Cuevas et al. 2019). The mathematical

---

A. Maheshwari (✉) · S. Jaiswal · Y. R. Sood · H. Raj
Department of Electrical Engineering, National Institute of Technology, Hamirpur, HP, India
e-mail: maheshwari.ankur@nith.ac.in

S. Sabyasachi
Yeungnam University, Gyeongsan, Korea

C. B. Khadse et al. (eds.), *Intelligent Methods in Electrical Power Systems*, Engineering Optimization: Methods and Applications, https://doi.org/10.1007/978-981-97-5718-3_1

formulation of the constrained optimization problem (OP) is presented as follows (Institution of Engineering and Technology 2018).

$$\text{Find } x = \begin{bmatrix} x_1 \\ x_2 \\ . \\ . \\ x_n \end{bmatrix} \text{ which minimize or maximize } F(x) \tag{1.1}$$

Subjected to

$$g(x) = 0 \tag{1.2}$$

$$h(x) \leq 0 \tag{1.3}$$

$$x \in X \tag{1.4}$$

where x denotes n-dimensional vector of decision or control variables, $g(x)$ and $h(x)$ is a vector of equality and inequality constraints, respectively and F represents the objective functions (OF) to minimize or maximize. The space of alternative solutions to a problem is represented by X, and the feasibility zone of the decision variables is defined by the condition (1.4). An unconstrained optimization problem contains no such constraints and can be defined as (1.1). The objective function $F(x)$ can be non-convex, non-linear, and non-differentiable in nature, and constraints $g(x)$ and $h(x)$ can be both continuous and discrete in nature. As a result, the optimization problem defined by (1.1)–(1.4) might be extremely challenging.

Optimization techniques from operations research are harnessed to determine the best solution for fixing the planning and operation problems today's modern power systems face. These optimization techniques have two primary classifications: (i) classical or exact optimization approaches and (ii) metaheuristic optimization approaches. The exact algorithms ensure the best solution in a finite period and must demonstrate that the acquired solution is optimal. Many classical algorithms are deterministic and employ knowledge of gradients. These algorithms rely on slopes such as the Newton-Raphson algorithm (Cuevas et al. 2020). These algorithms often solve smooth, unimodal problems using the fitness values of the underlying functions and their derivatives. Linear programming (LP), non-linear programming (NLP), dynamic programming (DP), mixed integer linear programming (MILP), and mixed integer non-linear programming (MINLP) are a few examples of classical optimization approaches. Unfortunately, these techniques are often stuck at local optima because of differentiability, non-linearity, and non-convex issues (Maheshwari et al. 2022a). In addition, the limitations of these approaches vary by problem type, for instance, when an algebraic representation of the objective function is unavailable. As a result, it is crucial to refine optimization techniques to conquer these drawbacks.

Therefore, several heuristic and metaheuristic approaches have been explored over the past few decades to solve complex restricted optimization issues.

Two terms are combined to form the "metaheuristic". The term "heuristic" refers to a method or technique that aids in the process of discovery. The other term, "meta," used as a prefix, indicates a high-level strategy directing the solution-finding process (Bandaru and Deb 2016). Converting models of physical processes or natural occurrences into computational tools is a central tenet of several metaheuristic algorithms (MA). In the past few years, metaheuristic optimization algorithms have developed as reliable tools for finding optimal solutions to challenging and complex power system problems. Such algorithm's computational efficiency and speed far outstrip those of derivative-based methods. The ability to avoid being stuck in local optimums gives these algorithms an advantage over gradient-based methods to obtain the near-global best solution (Carbas et al. 2021). As a result, many metaheuristic methods have been presented as viable replacements for classical optimization strategies. Most metaheuristics algorithm's results depend on the class of random variables generated because most algorithms use some form of stochastic optimization.

MA are classified into single-solution and population-based solution-based algorithms. In each iteration, the single solution-based algorithms produce a single solution in each iteration and solve optimization problems with less function evaluation (1 $\times$ *maximum iterations*). However, these algorithms are less likely to arrive at a global optimum. On the other hand, swarm-based optimization methods used several solutions during each iteration. These procedures begin by generating random solutions as the initial population, and then algorithm operators are used to improve the initial population by having them communicate the information between themselves. A potential drawback of swarm-based algorithms is the high number of function evaluations (*population size* $\times$ *maximum number of iterations*) (Karami et al. 2021). Despite this drawback, swarm-based algorithms have a better chance of discovering global optimum than single-based algorithms since they use a population and share information among individuals inside the population (Gavrilas 2010).

On the other hand, the optimization problem objectives can be categorized into single, multi, and many-objective optimization.

Single-objective Optimization: Single-objective optimization focuses solely on maximizing or minimizing the single objective function.

Multi-objective Optimization: When two or more objectives are minimized or maximized simultaneously, it presents a challenge known as multi-objective optimization (Bandyopadhyay and Saha 2013). Decision-making is greatly aided by multi-objective optimization methods, especially when there is a conflict between competing objectives. In this situation, a strategy built on the ideas of Pareto dominance can be helpful. This method considers a solution non-dominated when no other solution yields superior values for any objectives. The collection of non-dominant alternatives constitutes the Pareto front, which consists of compromise solutions from which the decision-maker can select the most desirable.

Many-objective Optimization: It's a special case of multi-objective optimization in which the number of goals is usually greater than two. Thus, multiple Pareto-optimal solutions are produced rather than a single optimal one. According to surveys, this is one of the most rapidly developing areas of computational intelligence. Research in this area has attracted attention from various disciplines, including mathematics, computer science, economics, and engineering (Li et al. 2015). Metaheuristics are becoming increasingly popular as a solution method for problems with multi and many objectives optimization problems.

In this study, an overview of metaheuristic algorithms with their practical application in power system optimization problem is addressed in Sect. 1.2. In Sect. 1.3, a comprehensive overview of widely used metaheuristic algorithms specifically employed in power system optimization is presented. Section 1.4 focuses on dispelling popular misconceptions prevalent in current literature and proposes viable solutions. The final section concludes the study with a concise summary and highlights the key findings.

1.2 Metaheuristic Algorithm Formulation

An effective metaheuristic algorithm (MA) should be able to explore the search space efficiently to find near-optimal or best solutions. These are based on a search process-driving technique that may draw inspiration from any observed artificial or natural system. This can stem from various sources, including the annealing process in the metallurgy or the activity of foraging ants.

There are two main features of any MA: exploration and exploitation. Exploration means how successfully operators broaden the range of possible solutions in the search space, which gives a global search ability to MA. Exploitation is the degree to which the operators utilize the information provided by the solutions in earlier iterations to speed up the search process, which enhances the local search ability of these algorithms (Stegherr et al. 2022). The advantages of metaheuristics over classical optimization methods can be summarized as follows:

1. MA can produce satisfactory solutions for relatively easy problems with significant input complexity, which might be a hurdle for classical approaches.
2. When no exact algorithm can solve a problem in a reasonable period, metaheuristics can nonetheless lead to satisfactory outcomes.
3. MA can be applied with black-box, non-analytic, or simulation-based objective functions, whereas most classical approaches require gradient knowledge.
4. The majority of metaheuristics can overcome the situation of local optima.
5. Due to inherent stochasticity or deterministic heuristics, most metaheuristics can recover from local optima.
6. With the ability to recover from local optimums, MA can better manage objective uncertainty.

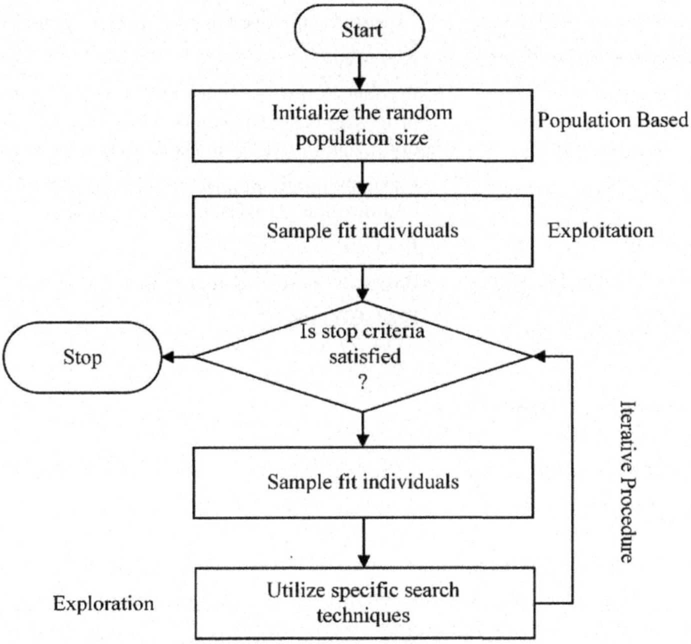

Fig. 1.1 The basic structure of the metaheuristic algorithm

7. Almost all metaheuristics can accommodate multiple objectives with minimal algorithmic modifications.

The majority of MA use fitness functions to evaluate potential solutions. This is to sample the most effective solutions to concentrate on exploitation. In addition, certain aspects of the search technique are incorporated to bring randomness and emphasize exploration. Since each search technique is different, it is challenging to represent this with a universal formulation (Santamaría et al. 2020). Figure 1.1 shows the basic structure of the metaheuristic algorithms.

1.2.1 Applications of Metaheuristic Algorithms (MA) in Power System Optimization Problems

This section addressed a few applications of power system problems typically addressed by metaheuristic optimization. Table 1.1 shows a few of the most popular metaheuristics used to solve these issues.

Table 1.1 Commonly used metaheuristic methods employed to solve power systems problems

Specific power system problems	MA used
Unit commitment	Particle swarm optimization (PSO), evolutionary programming (EP), genetic algorithm (GA), ant lion optimization (ALO), ant colony optimization (ACO)
Economic dispatch	Dynamic programming (DP), teaching learning-based optimization (TLBO), PSO, GA, ALO
Optimal power flow	PSO, EP, GA, ALO, TLBO
Forecasting of demand and supply	PSO, EA, simulated annealing (SA), TLBO, GA
Power system protection	PSO, GA, and SA

1.2.1.1 Unit Commitment (UC)

Daily and weekly operational planning for today's advanced power grids relies heavily on Unit Commitment. Electric utilities decide which generator to start, when to switch into the grid, in what sequence to shut them down, and for how long, all to optimize economic performance. The problem involves integer and continuous variables, including constraints like minimum up and down times, spinning reserve, controlled ramp rate, power balance, start-up and shutdown costs, and many more (Kumar et al. 2021b). There are several forms of unit commitment problems, such as conventional UC, security-constrained UC, and profit-based UC. Scheduling in a vertically integrated system and scheduling in a deregulated environment are two ways UC can be categorized from the perspective of market operations (Kumar et al. 2022).

DP, MILP, Lagrangian relaxation (LR), and priority listing (PL) are some of the conventional methods that have been used to address the UC problem (Håberg 2019). The PL method is advantageous over other approaches because of its simplicity and fast computational speed. However, it has high operating costs due to constrained exploration operations in the search space. DP requires a significant amount of computational time and suffers from the curse of dimensionality. Unfortunately, like many other approaches to solving large-scale problems, MILP has the drawback of requiring enormous computational time and ample storage space. Despite delivering quick solutions to UC problems, the LR method suffers from numerical convergence and suboptimal outcomes.

The above-mentioned approach's shortcomings prompted extensive study of metaheuristic optimization techniques, which in turn gave rise to optimization tools like the GA (Rodríguez del Nozal et al. 2020), PSO (Anand et al. 2019), Monarch Butterfly Optimization (MBO) (Kumar and Naresh 2020), ALO (Yasin et al. 2021), shuffled frog leaping algorithm (SFLA) (Ebrahimi et al. 2011), grey wolf optimizer (GWO) (Sakthi et al. 2017), EP (Jang et al. 2011), and whale optimization algorithm (WOA) (Reddy et al. 2019).

1.2.1.2 Economic Dispatch (ED)

After deciding the commitment of power generating units, the proceeding step is to determine the power needed to be generated by each generating unit to cater completely to the load demand and transmission network constraints so that overall generation cost is minimized (Kumar et al. 2021a). The valve-point loading effect in the input-output curve of the generation units is one source of non-linearity in the ED problem (Zou et al. 2019).

There is also the option of simultaneously considering the unit commitment and economic dispatch problem called Network-Constrained Unit Commitment (NCUC). One seeks to determine the intervals over which various generators are operational, and the other is to determine the power output of all such generators over the specified periods. Several constraints considered in the ED problem include equality constraints like the power balance equation, inequality constraints such as upper and lower bounds of power, multiple fuel inputs, prohibited operating zones, and transmission line limits (Chicco and Mazza 2020).

DP (Kumar and Palanisamy 2007), quadratic programming (Fan and Zhang 1998), and linear programming (Hoke et al. 2013) are some of the more standard classical algorithms employed to solve ED problems. Over the past decades, several MAs have been applied to the ED problem to overcome the difficulties introduced by the non-convexity of the domain of the definition of the variables. These include PSO (Tiwari et al. 2020), gravitational search algorithm (GSA) (Swain et al. 2012), GA (Jan 2021), ALO (Alazemi and Hatata 2019), and GWO (Pradhan et al. 2016). Hybridizations of mathematical programming and metaheuristics, such as GWO and PSO (Maheshwari and Sood 2022), have also been used to develop specialized solvers.

1.2.1.3 Optimal Power Flow (OPF)

Power system operators performed OPF for day-ahead markets, hourly, and even in 5–15 min (Maheshwari and Sood 2022). OPF is a mixed integer, non-convex, and non-linear optimization problem. It seeks to identify a system's optimal steady state operating point, considering both single and multi-objective formulations to be minimized or maximized (Maheshwari et al. 2022b). In other words, OPF optimizes a specified objective function by regulating power flow without going against power flow or operational constraints. OPF is an essential tool for the system operator in resolving numerous power system planning, management, and control-related issues. There have been several reported variants of the OPF, some of which are shown in Fig. 1.2.

Typically, OPF problems contain two variables: control or decision variables and state variables. The best value for the decision variables is primarily identified to satisfy numerous system and operational constraints. Then dependent variables should be calculated based on the best value for the decision variables (Abdi et al. 2017). It is essential to highlight that the number of decision variables determines the

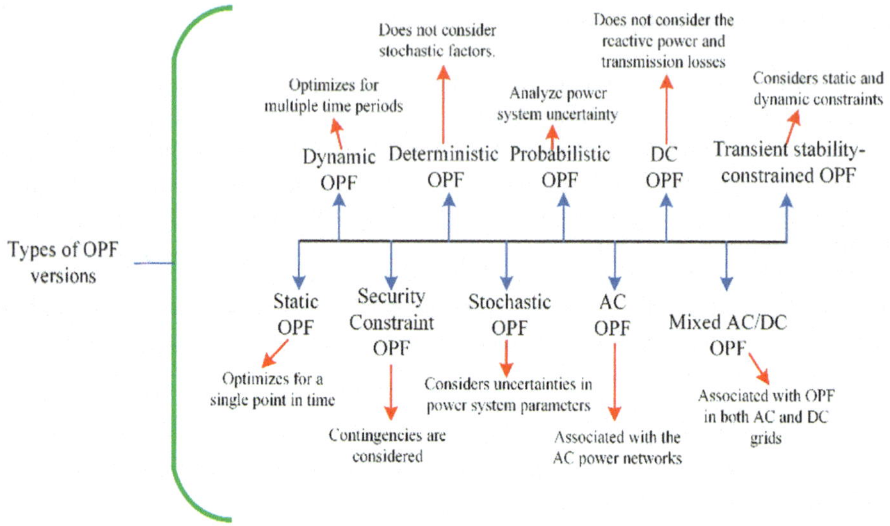

Fig. 1.2 A variety of OPF versions have been reported so far

solution space. Specifically, if there are n decision variables, there is n dimensional solution space.

The OPF problem, an incredibly advanced real-time OPF problem, is studied using various mathematical programming techniques. Recent years have seen the popularity of interior point techniques as a viable method for solving complex problems. But MA works well for the OPF problem and yields decent outcomes. GA (Osman et al. 2004), PSO (Maheshwari et al. 2022a), EA (Khamees et al. 2017), ALO (Maheshwari et al. 2021), the flow direction algorithm (Maheshwari et al. 2023a), and the flower pollination algorithm (Kathiravan and Kumudini Devi 2017) are some of the most well-known metaheuristics. Well-planned hybridizations of other algorithms, such as hybrid PSO-GWO (Riaz et al. 2021), also provide alternative options.

1.2.1.4 Forecasting of Demand and Supply

Predicting future electrical loads is an age-old challenge that has been addressed by a wide range of techniques, from simple statistical analysis to AI-based systems like neural networks and support vector machines and, more recently, deep learning.

Today's increased generation from renewable energy sources brings a great deal of uncertainty because it depends on variables such as solar irradiation, wind speed and direction, and power pricing. So, methods are required that can also tackle the generation forecasting problem. Since the time series for predicting load and generation are two different problems, each is often treated as independent of the other. Forecasting horizons are crucial for problem definition. From a classical perspective, seconds to minutes, hours to weeks, and months to years are all considered different time frames (from some months to many years).

Generalized standards are measured against persistence models, which are built by recreating the time series that is thought to be most predictive of the future. For quite some time, neural network and fuzzy system-based statistical methodologies and methods have been used. An algorithm that aids parameter adjustment during training has been integrated into neural networks to create hybrid approaches. These hybridizations have also taken MA into account.

PSO, TLBO, and ALO have all been employed, although GA is the most common. Combining diverse forecasting techniques, ensemble-based forecast models are becoming more widely used as useful tools.

1.2.1.5 Power System Protections (PSP)

Power system protection plays a crucial role in ensuring the safe and reliable operation of electrical power systems. Traditional protection schemes are based on deterministic approaches, which may not always yield optimal solutions in complex and uncertain operating conditions. On the other hand, Metaheuristic algorithms (MA) offer powerful optimization techniques that can effectively address these challenges. A few applications of metaheuristic algorithms in PSP are as follows.

Fault Detection and Classification: MA can effectively address fault detection and classification problems by optimizing the selection of appropriate features and classifier parameters. The algorithms can exploit the information contained in fault signatures, enabling accurate fault detection and classification, thus improving the reliability and response time of protection systems.

Optimal Coordination of Protective Devices: Determining the optimal coordination of protective devices, such as relays and circuit breakers, is critical to power system protection. MA can optimize the coordination settings by considering multiple objectives, such as minimizing the total operating time and the impact on system stability. This approach improves selectivity and coordination of protective devices while ensuring system reliability. Differential evolution, harmony search, and artificial bee colony algorithms have shown promise in achieving optimal relay coordination.

Protection Setting Optimization: Setting the protection parameters, such as relay settings, is challenging due to the complexity and uncertainties in power systems. MA can optimize the protection settings to achieve multiple objectives, including minimizing operating time, reducing coordination conflicts, and maximizing system stability. This approach enhances the overall performance and adaptability of protection systems.

Protective Device Placement: The optimal placement of protective devices is crucial for efficient fault detection and localization. MA can determine the optimal locations for protective devices, such as relays, by considering various factors, including system topology, load conditions, and fault statistics. This approach improves the coverage and sensitivity of protection systems. PSO, GA, and SA have been successfully employed for protective device placement optimization.

The application of MA in power system protections offers several advantages, including improved fault detection accuracy, optimized coordination schemes, enhanced reliability, and adaptive protection settings. However, challenges such as algorithm parameter selection, convergence speed, computational complexity, and algorithm robustness must be carefully addressed to ensure reliable and practical implementation.

1.3 Overview of Widely Used Metaheuristic Algorithms for Power System Optimization Problems (PSOP)

The commonly used metaheuristic algorithms for the above-discussed PSOP are discussed below.

1.3.1 Genetic Algorithm (GA)

GA was introduced in 1970 (Taylor 1994), which has since become widely utilized as an optimization tool in power systems optimization. Some evolutionary theorists have assumed that only the fittest and robust species will survive in nature, and this idea inspired this algorithm (Mahdi et al. 2018). The GA algorithm utilizes a set of chromosomes called population. It starts with arbitrary population initialization, iteratively seeks a better and better solution, and eventually settles on a single best solution (Abusorrah 2014). The generalized flowchart of GA is shown in Fig. 1.3.

1.3.2 Particle Swarm Optimization (PSO)

After years of research and development, Kennedy and Eberhart (Kennedy 2017) developed PSO, a powerful stochastic optimization approach. It explores the search space of a problem to determine the parameters needed to optimize the objective function. Swarm intelligence is the inspiration for PSO, which was used to create an algorithm for dealing with realistic stochastic issues by drawing on natural lessons (Gad 2022). The PSO algorithm has been effectively applied to many PSOP, and it is rising in popularity in computational intelligence.

The particles, or individuals in PSO, linked up, interact, and communicate reliably with each other by employing search gradients or directions. PSO employs established particles to swarm around a search space in pursuit of global optimums. Each

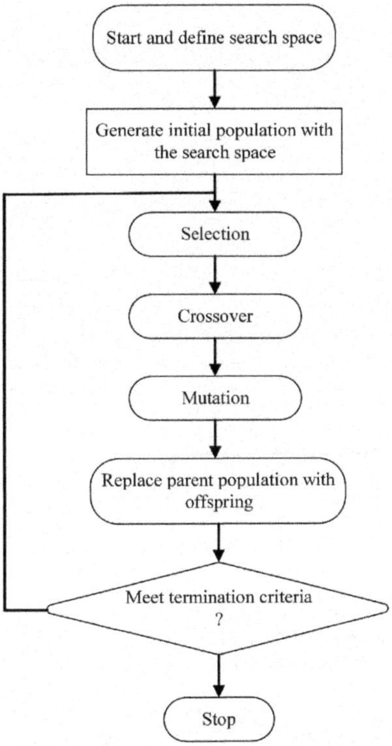

Fig. 1.3 Generalized flowchart of genetic algorithm

particle in the PSO updates its position throughout the iteration process using its current knowledge and the information gained from the neighborhood search. This means that when a group of birds, butterflies, or fish navigates, the group members keep their positions and gain knowledge from their collective experiences. Figure 1.4 depicts the generalized flowchart for PSO.

1.3.3 Ant Lion Optimization (ALP)

ALO is a novel optimization algorithm inspired by nature, first presented by Dorigo et al. (2006). ALO algorithm simulates the foraging behavior of ant lions, and the algorithm is developed in five stages, each inspired by the hunting strategy of an ant lion. The ant lion employs five primary methods of hunting. The first approach starts with the simulation of a random walk of ants. The second process is related to constructing traps for the prey. The third method is the technique of stinging and entrapping ants in traps. The fourth step involves repositioning or securing the prey, and the last step consists in reassembling the traps (Saini and Rahi 2022). By modeling the ALO‘s actual behavior, precise and accurate mathematical equations of the ALO have been developed. This can be applied to many different real-world

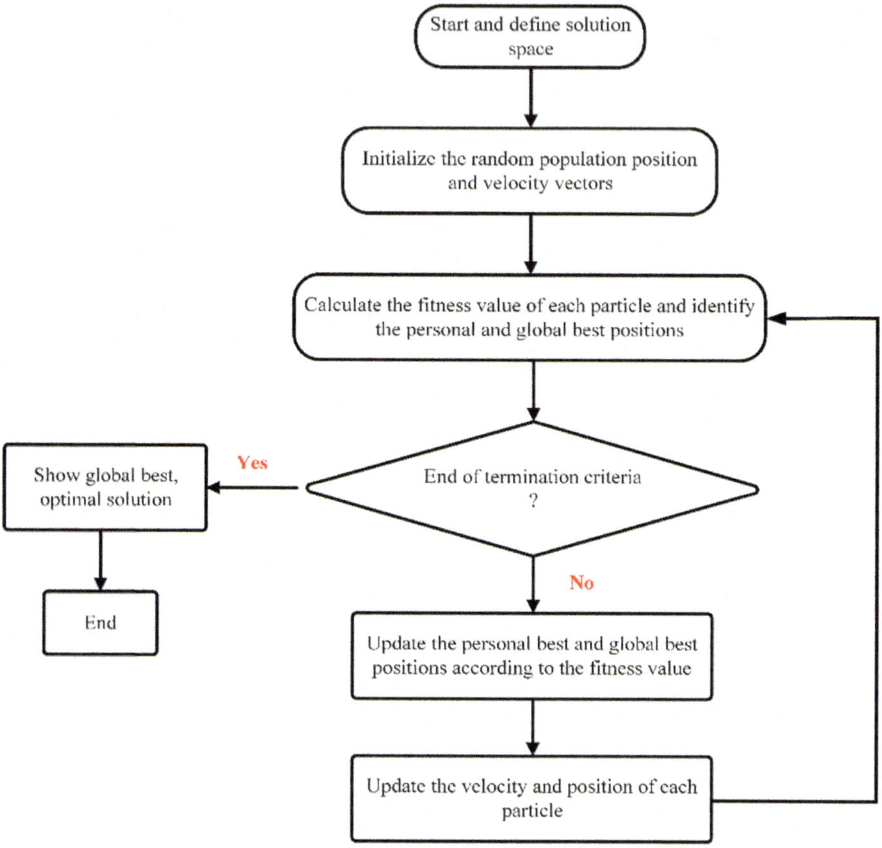

Fig. 1.4 Generalized flowchart of PSO

power system problems. The generalized flow chart of the ALO algorithm is depicted in Fig. 1.5.

1.3.4 Grey Wolf Optimization (GWO)

GWO is a population-based metaheuristic algorithm. The development of the algorithm was inspired by the techniques used by Grey Wolves during their hunts (Moazami Goodarzi and Kazemi 2018). In GWO, each pack member is placed in one of four distinct echelons, labeled from highest to lowest, as shown in Fig. 1.6. These echelons are designated alpha (α), beta (β), delta (δ), and omega (ω). In this context, α represents the top individual in charge of the park and is exemplified by authoritative decrees and decisions (Yu et al. 2023). Beta, on the other hand, represents an individual of the system who assists α during the park selection process

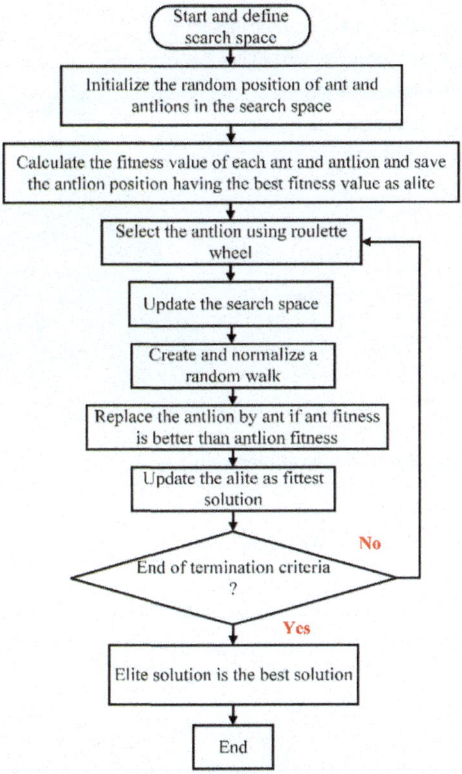

Fig. 1.5 Generalized flowchart of ALO algorithm

and participates in other activities. Beta issues orders to lower-ranking individuals and follows alpha's directives. Beta can play the role of a wise advisor to alpha. When it comes to park discipline, beta is on par with alpha. The ω on the pyramid represents the lowest-ranking members. Omega is the most faithful member of the group (Mishra 2022). The wolf pack's activities, including fighting, hunting, and cub-caring, could be vulnerable in the absence of ω. The delta is unique compared to the other three wolf packs on the pyramid's echelons. It's also worth noting that the wolves known as deltas are excessively subservient and loyal to the wolves known as α and β, while trying to exert dominance over the ω. They commonly act as a detective, lookouts, guards, hoariers, predators, or curators. A schematic of the GWO optimization method is depicted in Fig. 1.7.

1.3.5 Teacher Learning-Based Algorithm (TLBO)

TLBO was developed in 2011 by Rao et al. (2011). TLBO has two stages based on the classical teacher-student interaction: (i) the teaching phase and (ii) the learning

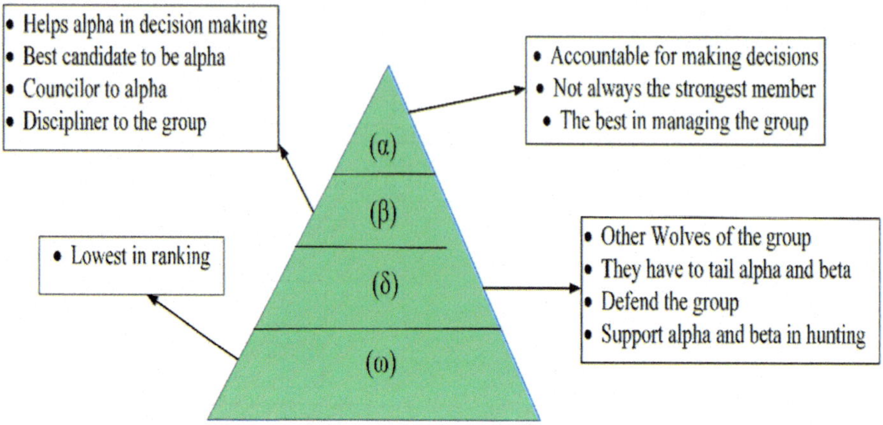

Fig. 1.6 Grey wolf hierarchy

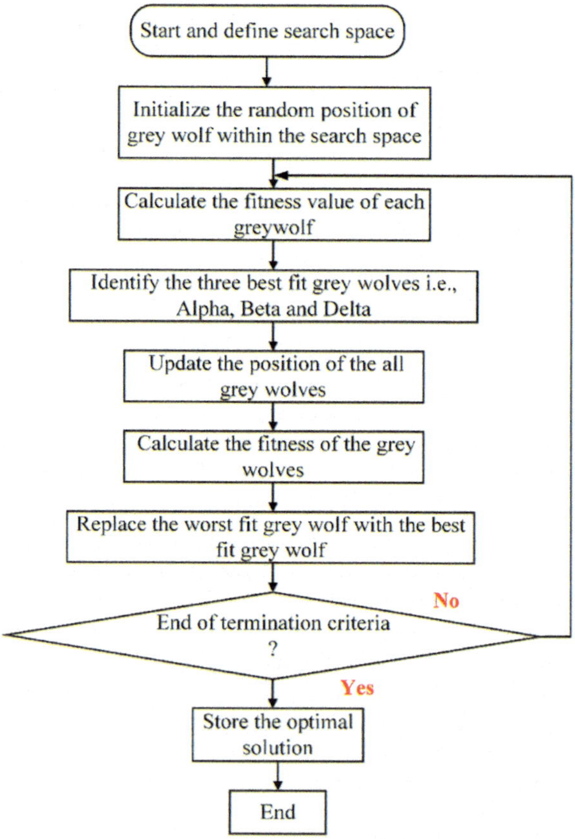

Fig. 1.7 Generalized flowchart of GWO algorithm

phase. TLBO's simple structure, ease of implementation, ability to deliver faster convergence, and the absence of algorithm-specific parameters are a few reasons why it's garnered increasing attention in recent years. In TLBO, the population is compared to a classroom full of students and the various control variables to a selection of courses (Bahmani-Firuzi and Khorshidi 2017). A student's performance is like the fitness function in a problem. This optimization method is meant to function similarly to a classroom, where a teacher with extensive expertise strives to raise his students' general level of understanding (Rao et al. 2012). In addition to interacting with others, students can communicate with themselves to increase their understanding. The end outcome will serve as the benchmark against which their expertise may be evaluated. The TLBO process flow is displayed in Fig. 1.8.

1.4 Potential Issues and Unsuitable Remarks Concerning the Effectiveness of Metaheuristic-Based Optimization

'Several authors make misleading claims about the efficacy of the methodologies utilized in their studies that seek to use metaheuristic optimization techniques. Many papers submitted to scientific publications or conferences are rejected for this reason. In this section, the most noteworthy (and often frequent) circumstances are recalled, along with considering whether more acceptable remedies could be implemented (Maheshwari et al. 2023b).

1.4.1 Global Optimum Achieved

Some research articles ascertain that "the global optimum" is the objective of the study. This statement is always false for any heuristic solution to a practical problem of any size. According to its findings, no metaheuristic technique can guarantee finding the global optimum in a specific amount of time using a predetermined number of repetitions. Only a small group of metaheuristics, those needing an infinite number of iterations (NOI), have had their asymptotic convergence to the global optimum illustrated. As is required for engineering problems, no practical heuristic can promise that the global optimum may be obtained in a definite amount of time.

1.4.2 Termination Criteria

A flexible termination criterion in the execution of MA is crucial. Surprisingly, many of the MA employed in existing articles only use the maximum NOI. Nonetheless,

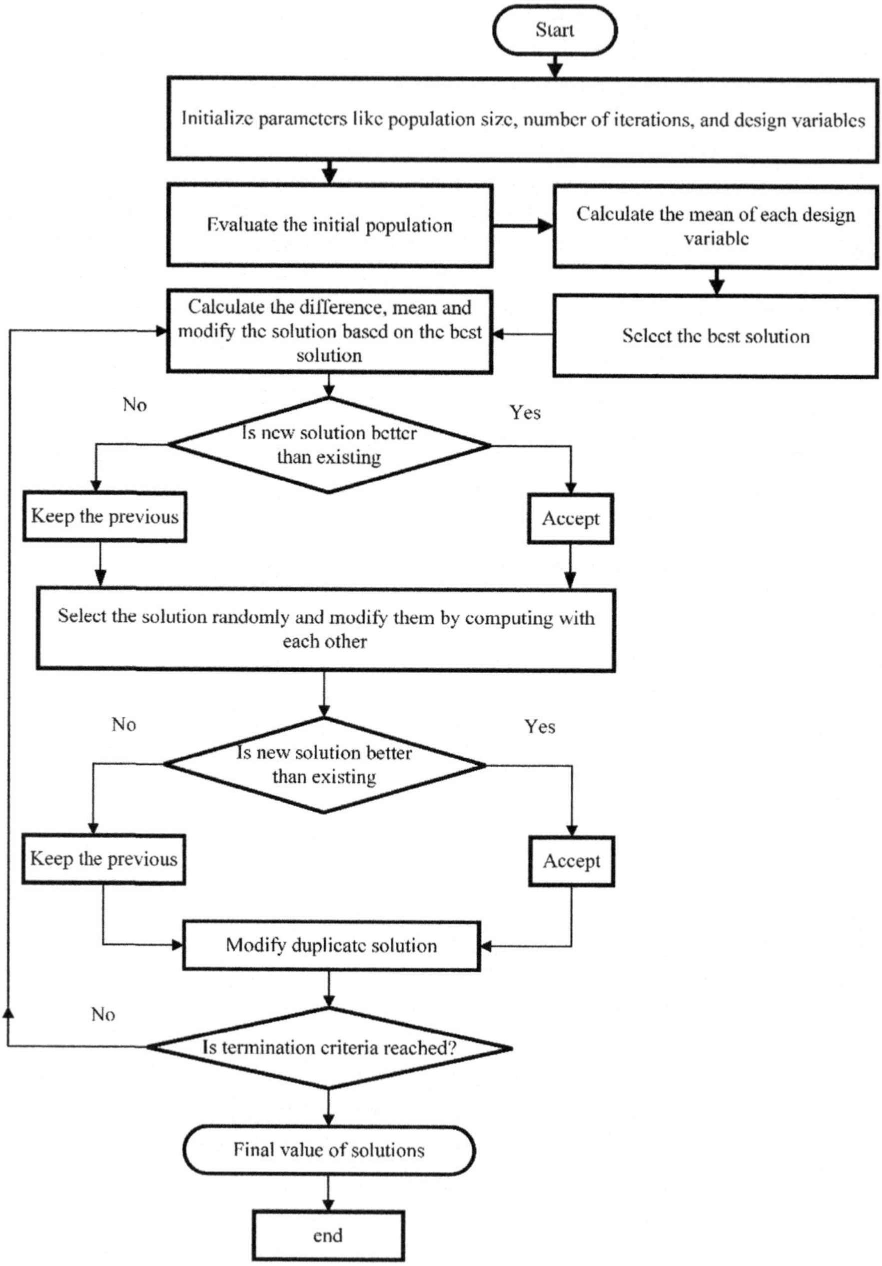

Fig. 1.8 Generalized flowchart of TLBO algorithm

in most cases, this selection would be wrong. Indeed, early and late termination are separate issues that can develop.

In the case of early termination, the progression of the objective function is still yielding substantial improvements. Late termination occurs when the execution of an algorithm is stopped after a large NOI without a meaningful change to the objective function. As a result, a lot of unnecessary constant values have been generated at the end of the execution and might have been removed.

There is always the chance that the considered objective function will get better during its progression. Therefore, it is crucial to develop a reliable termination criterion to use computational resources effectively, some of which are listed below.

- Once a certain NOI has passed with no improvement to the best objective function found so far, the algorithm stops. Both premature termination and overdue termination are prevented here.
- The user can adjust the NOI based on their knowledge of how the objective function changes over time for their problem of interest. Because of this, a very high setting for the maximum NOI may be the only way to effectively prevent the possibility of infinite repetitions of the same operation.
- The computation time could be used as the stop condition (which is the same for all methods) if the MA is performed to compare performance. More importantly, the maximum number of allowed iterations is not required here. For this reason, stopping the execution of a MA based on the maximum number of allowed iterations is not a good idea.

1.4.3 Not Only Metaheuristics

Many published works show that fundamental problems (convex in nature) may be handled with the help of metaheuristics, which is not a contribution. Researchers need to be more explicit about why they are considering using heuristic algorithms. Heuristic procedures are only recommended by researchers when empirical methods have proven unsuccessful. Analytical approaches are preferable because there are several solvers accessible nowadays. A heuristic algorithm will be considered if these solvers are proven ineffective.

1.4.4 The Significance of Fast Convergence

When demonstrating the "superiority" of the proposed algorithm over other methods, several contributions emphasize the significance of obtaining fast convergence. In this context, a statement like "The convergence of the proposed algorithm is better than for other approaches" is given. This means that the objective function of the proposed algorithm improves in a lesser NOI than it does for existing approaches.

It is important to remember that the time required to solve a single iteration might vary between algorithms. Therefore, it is meaningless to evaluate the NOI in isolation. In addition, the progress obtained with fewer iterations may not be significant. Fast convergence can typically be achieved by making an initial guess for single-update methods or good initial population selections for population-based methods. Convergence to the optimal solution is immediate if the initial guess already includes the global optimum. This does not indicate that the selected algorithm is superior to others.

1.4.5 The Number of Individual Runs

The metaheuristic approach is often only used for fewer individual runs, a common shortcoming of many articles. A significant number of individual runs is required to achieve statistical significance for the selected algorithm, preferably more than 100. Of course, it would be ideal if the problem could be solved hundreds of times in a fair amount of computational time. Similarly, if the iterative procedure has only been executed once for each technique, then a comparison among different algorithms is relatively meaningless.

1.4.6 Eliminating Misconceptions

One potential concern that may arise while evaluating a MA is that "the results achieved are substantially superior to those available in the literature". What might have happened is that the proposed algorithm has yielded significantly better results than the results of similar problems found in the literature using other algorithms. The warning for this scenario is that the outcome may be questionable. Potential problems with the underlying modeling may be the source of the current predicament. For instance, it might be possible that the system under consideration differs slightly from that of the other sources. Alternatively, the constraints may vary slightly among the studied approaches. Solving the problem on a small test system where the global optimum is known might provide helpful insight into this situation; if the results are different, it suggests that either the data or the algorithm are flawed.

1.5 Conclusion

This study represented a foundational understanding of the most prevalent metaheuristic algorithms and their application to typical optimization issues in power systems. This study concludes that metaheuristic algorithms are the most effective for handling challenging power system optimization problems, and published works on metaheuristic applications in the power systems domain are expanding at an astounding rate. The general description of various renowned and encouraging methodologies has been presented with flowcharts. In addition, some frequent errors in published publications are emphasized. In this context, the following recommendations have been proposed for creating meaningful contributions to the application of metaheuristic algorithms to power system problems.

- Use flexible termination criterion as the primary for the proposed algorithm. The termination of algorithm execution based on the maximum NOI should be considered secondary.
- In order to prevent differences in data and problem definition between articles, it is best to implement and execute the comparative algorithms using the same data and problem formulation.
- A substantial amount of executions (not just one) and the application of suitably reliable statistical indicators were used to determine the statistics of the outcomes achieved on standard or practical test systems.
- Avoid incorrect terminology and prematurely declaring the effectiveness of an algorithm based on results acquired on a single problem and with insufficient testing.

References

A review on the unit commitment problem: approaches, techniques, and resolution methods

Abdi H, Beigvand SD, Scala ML (2017) A review of optimal power flow studies applied to smart grids and microgrids. Renew Sustain Energy Rev 71:742–766

Abusorrah AM (2014) The application of the linear adaptive genetic algorithm to optimal power flow problem. Arab J Sci Eng 39(6):4901–4909

Alazemi FZ, Hatata AY (2019) Ant lion optimizer for optimum economic dispatch considering demand response as a visual power plant. Electr Power Compon Syst 47(6–7):629–643

Anand H, Narang N, Dhillon JS (2019) Multi-objective combined heat and power unit commitment using particle swarm optimization. Energy (Oxf) 172:794–807

Bahmani-Firuzi B, Khorshidi R (2017) A new optimization algorithm based on teacher learning algorithm for optimal operation of electric grids. J Intell Fuzzy Syst 32(6):3839–3846

Bandaru S, Deb K. (2016) Metaheuristic techniques. In: Decision sciences. CRC Press, Taylor & Francis Group, 6000 Broken Sound Parkway NW, Suite 300, Boca Raton, pp 693–750

Bandyopadhyay S, Saha S (2013) Some single- and multi-objective optimization techniques. In: Unsupervised classification. Springer Berlin Heidelberg, Berlin, Heidelberg, pp 17–58

Carbas S, Toktas A, Ustun D (2021) Introduction and overview: nature-inspired metaheuristic algorithms for engineering optimization applications. In: Springer tracts in nature-inspired computing. Springer Singapore, Singapore, pp 1–9

Chicco G, Mazza A (2020) Metaheuristic optimization of power and energy systems: underlying principles and main issues of the 'rush to heuristics.' Energies 13(19):5097

Cuevas E, Barocio Espejo E, Conde Enríquez A (2019) Introduction to metaheuristics methods. In: Metaheuristics algorithms in power systems. Springer International Publishing, Cham, pp 1–8

Cuevas E, Fausto F, González A (2020) An introduction to nature-inspired metaheuristics and swarm methods. In: Intelligent systems reference library. Springer International Publishing, Cham, pp 1–41

Dorigo M, Birattari M, Stutzle T (2006) Ant colony optimization. IEEE Comput Intell Mag 1(4):28–39

Ebrahimi J, Hosseinian SH, Gharehpetian GB (2011) Unit commitment problem solution using shuffled frog leaping algorithm. IEEE Trans Power Syst 26(2):573–581

Fan J-Y, Zhang L (1998) Real-time economic dispatch with line flow and emission constraints using quadratic programming. IEEE Trans Power Syst 13(2):320–325

Gad AG (2022) Particle swarm optimization algorithm and its applications: a systematic review. Arch Comput Methods Eng 29(5):2531–2561

Gavrilas M (2010) Heuristic and metaheuristic optimization techniques with application to power systems. D. Mangeron Blvd

Håberg M (2019) Fundamentals and recent developments in stochastic unit commitment. Int J Electr Power Energy Syst 109:38–48

Hoke A, Brissette A, Chandler S, Pratt A, Maksimovizc D (2013) Look-ahead economic dispatch of microgrids with energy storage, using linear programming. In: 2013 1st IEEE conference on technologies for sustainability (SusTech)

Institution of Engineering and Technology (2018) Overview of metaheuristic optimization. In: Metaheuristic optimization in power engineering, Institution of Engineering and Technology, pp 1–38

Jan ZN (2021) Economic load dispatch using Lambda iteration, particle swarm optimization and genetic algorithm. Int J Res Appl Sci Eng Technol 9(8):972–977

Jang S-H, Roh J-H, Kim W, Sherpa T, Kim J-H, Park J-B (2011) A novel binary ant colony optimization: application to the unit commitment problem of power systems. J Electr Eng Technol 6(2):174–181

Karami H, Anaraki MV, Farzin S, Mirjalili S (2021) Flow direction algorithm (FDA): a novel optimization approach for solving optimization problems. Comput Ind Eng 156(107224):107224

Kathiravan R, Kumudini Devi RP (2017) Optimal power flow model incorporating wind, solar, and bundled solar-thermal power in the restructured Indian power system. Int J Green Energy 14(11):934–950

Kennedy J (2017) Particle swarm optimization. In: Encyclopedia of machine learning and data mining. Springer US, Boston, MA, pp 967–972

Khamees AK, El-Rafei A, Badra NM, Abdelaziz AY (2017) Solution of optimal power flow using evolutionary-based algorithms. Int J Eng Sci Technol B, Lagos 9(1):55

Kumar V, Naresh R (2020) Monarch butterfly optimization-based computational methodology for unit commitment problem. Electr Power Compon Syst 48(19–20):2181–2194

Kumar SS, Palanisamy V (2007) A dynamic programming based fast computation Hopfield neural network for unit commitment and economic dispatch. Electric Power Syst Res 77(8):917–925

Kumar S, Kumar V, Katal N, Singh SK, Sharma S, Singh P (2021a) Multiarea economic dispatch using evolutionary algorithms. Math Probl Eng 2021:1–14

Kumar V, Naresh R, Singh A (2021b) Investigation of solution techniques of unit commitment problems: a review. Wind Eng 45(6):1689–1713

Kumar V, Naresh R, Sharma V (2022) Profit based unit commitment problem solution using metaheuristic optimisation approach. Int J Syst Sci Oper Logist 1–22

Li K, Deb K, Zhang Q, Kwong S (2015) An evolutionary many-objective optimization algorithm based on dominance and decomposition. IEEE Trans Evol Comput 19(5):694–716

Mahdi FP, Vasant P, Kallimani V, Watada J, Fai PYS, Abdullah-Al-Wadud M (2018) A holistic review on optimization strategies for combined economic emission dispatch problem. Renew Sustain Energy Rev 81:3006–3020

Maheshwari A, Sood YR (2022) Solution approach for optimal power flow considering wind turbine and environmental emissions. Wind Eng 46(2):480–502

Maheshwari A, Sood YR, Jaiswal S, Sharma S, Kaur J (2021) Ant lion optimization based OPF solution incorporating wind turbines and carbon emissions. In: 2021 Innovations in power and advanced computing technologies (i-PACT)

Maheshwari A, Sood YR, Sharma S, Sharma NK (2022a) Optimal power flow and its vindication in deregulated power sector. In: Lecture notes in electrical engineering. Springer Singapore, Singapore, pp 173–182

Maheshwari A, Sood YR, Jaiswal S (2022b) Investigation of optimal power flow solution techniques considering stochastic renewable energy sources: review and analysis. Wind Eng 0309524X2211240

Maheshwari A, Sood YR, Jaiswal S (2023a) Flow direction algorithm-based optimal power flow analysis in the presence of stochastic renewable energy sources. Electric Power Syst Res 216(109087):109087

Maheshwari A, Sood YR, Jaiswal S (2023b) A comprehensive review on stochastic optimal power flow problems and solution methodologies. IETE Tech Rev 1–28

Mishra KK (2022) Grey wolf optimization. In: Nature-inspired algorithms. CRC Press, Boca Raton, pp 131–143

Moazami Goodarzi H, Kazemi MH (2018) An optimal autonomous microgrid cluster based on distributed generation droop parameter optimization and renewable energy sources using an improved grey wolf optimizer. Eng Optim 50(5):819–839

Osman MS, Abo-Sinna MA, Mousa AA (2004) A solution to the optimal power flow using genetic algorithm. Appl Math Comput 155(2):391–405

Pradhan M, Roy PK, Pal T (2016) Grey wolf optimization applied to economic load dispatch problems. Int J Electr Power Energy Syst 83:325–334

Rao RV, Savsani VJ, Vakharia DP (2011) Teaching–learning-based optimization: a novel method for constrained mechanical design optimization problems. Comput Aided Des 43(3):303–315

Rao RV, Savsani VJ, Vakharia DP (2012) Teaching–learning-based optimization: an optimization method for continuous non-linear large scale problems. Inf Sci (NY) 183(1):1–15

Reddy KS, Panwar L, Panigrahi BK, Kumar R (2019) Binary whale optimization algorithm: a new metaheuristic approach for profit-based unit commitment problems in competitive electricity markets. Eng Optim 51(3):369–389

Riaz M, Hanif A, Hussain SJ, Memon MI, Ali MU, Zafar A (2021) An optimization-based strategy for solving optimal power flow problems in a power system integrated with stochastic solar and wind power energy. Appl Sci (Basel) 11(15):6883

Rodríguez del Nozal A, Tapia A, Alvarado-Barrios L, Reina DG (2020) Application of genetic algorithms for unit commitment and economic dispatch problems in microgrids. In: Nature inspired computing for data science. Springer International Publishing, Cham, pp 139–167

Saini A, Rahi OP (2022) Improved ant lion algorithm for solution of optimal power flow. In: 2022 3rd international conference on electronics and sustainable communication systems (ICESC)

Sakthi SS, Santhi RK, Murali Krishnan N, Ganesan S, Subramanian S (2017) Wind integrated thermal unit commitment solution using grey wolf optimizer. Int J Electr Comput Eng (IJECE) 7(5):2309

Santamaría J, Rivero-Cejudo ML, Martos-Fernández MA, Roca F (2020) An overview on the latest nature-inspired and metaheuristics-based image registration algorithms. Appl Sci (Basel) 10(6):1928

Stegherr H, Heider M, Hähner J (2022) Classifying metaheuristics: towards a unified multi-level classification system. Nat Comput 21(2):155–171

Swain RK, Sahu NC, Hota PK (2012) Gravitational search algorithm for optimal economic dispatch. Procedia Technol 6:411–419

Taylor CE (1994) Adaptation in natural and artificial systems: an introductory analysis with applications to biology, control, and artificial intelligence. Complex adaptive systems. John H. Holland. Q Rev Biol 69(1):88–89

Tiwari S, Pal NS, Ansari MA, Yadav D, Singh N (2020) Economic load dispatch using PSO. In: Micro-electronics and telecommunication engineering. Springer Singapore, Singapore, pp 51–64

Yasin ZM, Sam'on IN, Salim NA, Ab Aziz NF, Zakaria Z (2021) Optimal unit commitment for minimizing total operating cost using ant lion optimizer. In: 2021 IEEE international conference in power engineering application (ICPEA)

Yu X, Zhao Q, Lin Q, Wang T (2023) A grey wolf optimizer-based chaotic gravitational search algorithm for global optimization. J Supercomput 79(3):2691–2739

Zou D, Li S, Kong X, Ouyang H, Li Z (2019) Solving the combined heat and power economic dispatch problems by an improved genetic algorithm and a new constraint handling strategy. Appl Energy 237:646–670

Chapter 2
An Application of Artificial Bee Colony and Cohort Intelligence in the Automatic Generation Control of Thermal Power Systems

Murugesan Dhasagounder, Jagatheesan Kaliannan, Pritesh Shah, and Ravi Sekhar

Abstract The interconnected power system encompasses multiple regions, and various parameters play a crucial role in achieving stability and efficiency in power systems. Recent research has proposed a numerical optimization approach, specifically utilizing the Artificial Bee Colony (ABC) and Cohort Intelligence Optimization (CIO) methods, to fine-tune the parameters of a Proportional-Integral-Derivative (PID) controller for automatic generation control in a mutually dependent two-area power system with reheat thermal power generation. Both ABC and CIO are employed to determine the optimal gain values for the controller, involving the evaluation of multiple cost functions. To evaluate the effectiveness of these techniques in the context of the interconnected reheat thermal power system, a comparison is made between the performance of ABC-optimized controllers and CIO-optimized PID controllers. The findings of the study indicate that PID controllers optimized using the CIO method outperform those optimized using the ABC method in the context of the mutually dependent reheat thermal power system.

Keywords Artificial bee colony optimization · Cohort intelligence optimization · PID controller · Interconnected system · Automatic generation control

---

M. Dhasagounder (✉)
Paavai Engineering College, Namakkal, Tamil Nadu 637018, India
e-mail: benignsofts@gmail.com

J. Kaliannan
M. Kumarasamy College of Engineering, Karur 639113, India

P. Shah · R. Sekhar
Symbiosis Institute of Technology (SIT) Pune Campus, Symbiosis International (Deemed University) (SIU), Pune 412115, India

C. B. Khadse et al. (eds.), *Intelligent Methods in Electrical Power Systems*, Engineering Optimization: Methods and Applications, https://doi.org/10.1007/978-981-97-5718-3_2

2.1 Introduction

In the past, the electrical power grid was relatively uncomplicated and lacked complexity. Consequently, there was less emphasis on ensuring the stability of the network. However, as industries have grown and populations have expanded, it has become evident that these developments will lead to significant fluctuations in power system parameters, particularly in terms of frequency and voltage. Under normal load conditions, individual power plants autonomously maintain their stability and operational settings. However, when there is a sudden increase in load demand, it can disrupt the supply frequency and the power flow within the tie line connecting interconnected power systems (Elgerd and Fosha 1970; Kundur and Malik 2022; Jaleeli et al. 1992). Automatic generation control plays a crucial role in managing and regulating multiple interconnected systems, including solar and wind power generation, as well as electric vehicles. In this context, various controllers are employed as secondary control mechanisms within the system, with the goal of not only maintaining the system's frequency at a predetermined value but also responding to frequency changes as quickly as possible.

In reference Qian et al. (2013), an examination is conducted into the utilization of a neural sliding-mode approach, employing the Lyapunov method, for addressing the Load Frequency Control (LFC) problem. In Aldeen and Trinh (1994) explores the application of state feedback optimal control in a two-area integrated power system. The research investigates Load Frequency Control (LFC) with a PID controller tuned using an artificial neural network, taking into consideration the presence of distributed generation (Debnath et al. 2020). Reference Jood et al. (2019) discusses the design and simulation of an intelligent controller based on neuro-fuzzy techniques for a two-area power system with significant renewable energy integration. A simulation of the hydro-thermal area has been conducted using Matlab, and the study presents results for the application of fuzzy logic controllers in frequency stability control, particularly in the context of cross and tandem turbines, as described in reference Moniya and Anand (2021). Fractional order controllers applied to the load frequency control of microgrid power systems are discussed in reference Murugesan et al. (2023a). From the different control methods have certain drawbacks in the design process. For instance, in the case of multi-valued logic systems like fuzzy logic, neural networks, and ANFIS (Adaptive Neuro-Fuzzy Inference System), the design requires a control engineer with proper training to create fuzzy implication rules and process the controllers effectively. Sliding Mode Control can encounter noise-related issues. Conversely, designing a PID regulator/controller is relatively straightforward and user-friendly. It is easy to configure, operate, and maintain. Therefore, the PID regulator is generally considered the optimal choice for control engineers in practical applications.

In reference Murugesan et al. (2023b), an assessment is made of the performance of a single-area nuclear power plant's controller. The evaluation considers controller tuning methods, including Ziegler–Nichols, classical methods, and trial-and-error approaches, in addition to incorporating Cohort Intelligence algorithms.

In Pappachen and Fathima (2017), research has been conducted on load frequency control within a distributed generation power system, employing a range of control techniques and algorithms. The research conducted by Murugesan et al. (2022) involves an examination of the implementation of a cascaded PID (1 + PID) controller in both isolated and integrated power systems, considering the presence of uncertainties. References Shankar et al. (2017) and Arora et al. (2019) provide a comprehensive review of the studies and developments related to load frequency control and automatic generation control. The Authors (Behera et al. 2015) have examined, and several optimization algorithms have been applied to adjust the gain values of a PID regulator in the context of wind energy systems. The dynamic nature of the environment leads to variations in system specifications, which, in turn, impact the performance of the controller. Consequently, it becomes imperative for the optimal regulator gain values to be robust enough to safeguard against these changes in system parameters. In Murugesan et al. (2021), Ant Colony Optimization (ACO) was employed to assess the significance of a Proportional-Integral-Derivative with Derivative Filter (PIDD) regulator in comparison to a standard PID regulator for two-area thermal power networks integrated with renewable energy sources. In reference Mohanty et al. (2014), the Differential Evolution algorithm is employed to evaluate the significance of a PID compared to optimal feedback control in a triple-area multi-source power plant. Daneshfar and Bevrani (2012) have addressed the utilization of a Genetic Algorithm for H-infinity control design and a trial-and-error approach based on PI regulator tuning within a multi-objective optimization framework. The design of a PID regulator for the automatic generation control of double area power scheme is explored in reference Madasu et al. (2018), incorporating a multi-objective function and utilizing the Flower Pollination algorithm. A fuzzy PID controller optimized using the Firefly algorithm is introduced for the Automatic Generation Control (AGC) system. This controller is designed to operate in the presence of a Unified Power Flow Controller (UPFC) and Superconducting Magnetic Energy Storage (SMES) systems, and it is compared to both optimal control and the Differential Evolution algorithm in Pradhan et al. (2016). In reference Murugesan et al. (2023c), a comparative analysis is carried out to assess the performance of biological process-inspired algorithms like Genetic Algorithm (GA) and Differential Evolution (DE) in contrast to nature behavior-inspired algorithms such as Teaching–Learning-Based Optimization (TLBO) and Cohort Intelligence. This analysis focuses on tuning a PID controller within a single-area multi-source power system scheme.

All of the previously mentioned methods have been effectively employed, and advancements in system implementation have been assessed. However, certain algorithms require more programmer-defined parameters, which significantly impact the performance of the computational approach. Some of these algorithms exhibit longer processing times, involve complex execution procedures, lack a guaranteed solution, and may not exhibit a monotonic convergence rate. In contrast, the Cohort Intelligence Optimization algorithm (CIO), as introduced and refined by Kulkarni et al. (2013, 2019), is a newly developed algorithm that has proven successful in addressing real-time challenges and various engineering problems. This technique necessitates minimal user-defined parameters in the optimization process, demands very little data processing time, and is straightforward to apply and implement.

2.1.1 Organization of the Chapter

Following the outlined structure, the chapter is organized as follows: Sect. 2.1 provides an introduction and literature review. Section 2.2 delves into the development of the power system model, detailing the functional blocks and controller design. Section 2.3 explores swarm intelligence behavior and presents the flowchart for tuning the secondary controller. Section 2.4 is dedicated to the performance analysis of the two-area reheat thermal power system, including sensitivity and robustness analysis. Section 2.5 concludes the chapter by summarizing the key findings and insights.

2.1.2 Contribution of the Work

- A reheat thermal power generation system consisting of two interconnected areas has been established.
- A proportional–integral–derivative (PID) controller has been designed to implement the Automatic Generation Control (AGC) scheme within the proposed system.
- Cohort intelligence algorithms have been explored for the purpose of fine-tuning PID controllers in a two-area interconnected system. In order to demonstrate the superior performance of these proposed algorithms, the system's performance is compared with that of a PID controller optimized using the ABC algorithm.
- The dynamic response of the power system is analyzed for different initial population sizes in the algorithmic process, revealing improved performance with the PID controller.
- Additionally, a robustness and sensitivity analysis is carried out using the algorithm that yielded the best cost function results, considering variations in parameters and random load conditions.

2.2 Two Area Power System Model

A reliable power system simulation model mentioned in Fig. 2.1 considered for analysis is comprised of thermal and thermal power systems. PID regulators are independently advised for each area. In this article, different controllers are examined for areas 1 and 2. The mathematical expression of the PID regulator is mentioned in Eq. (2.1), and for optimizing gain values with ABC and Cohort algorithms, the objective functions considered are mentioned in Eqs. (2.2)–(2.5). The general block diagram representation of the PID regulator is mentioned in Fig. 2.2.

$$\text{Control signal} = \mathrm{U_{PID}} = \mathrm{K_P} + \frac{\mathrm{K_I}}{\mathrm{S}} + \mathrm{K_D s} \tag{2.1}$$

$$\mathrm{ACE_1} = \mathrm{ITAE} = \int_0^t t \cdot (|\Delta f_1| + |\Delta f_2| + |\Delta P_{tie}|)dt \tag{2.2}$$

$$\mathrm{ACE_2} = \mathrm{IAE} = \int_0^t (|\Delta f_1| + |\Delta f_2| + |\Delta P_{tie}|)dt \tag{2.3}$$

$$\mathrm{ACE_3} = \mathrm{ITSE} = \int_0^t \left(\left(t \cdot (\Delta f_1)^2 dt\right) + \left(t \cdot (\Delta f_2)^2 dt\right) + \left(t \cdot (\Delta f_{tie12})^2 dt\right)\right) \tag{2.4}$$

$$\mathrm{ACE_4} = \mathrm{ISE} = \int_0^t \left((\Delta f_1)^2 dt\right) + \left((\Delta f_2)^2 dt\right) + \left((\Delta f_{tie12})^2 dt\right) \tag{2.5}$$

where t-simulation time, Δf_1, and Δf_2 are the frequency inconsistency in Hz for areas 1 and 2 respectively, ΔP_{tie} represents a change in power in Mega Watts. $\mathrm{K_P}$, $\mathrm{K_I}$, and $\mathrm{K_D}$ are the tuning parameters of the PID regulator, i.e., Proportional, Integral, and Differential gain respectively. ACE is the Area control Error, which is the problem to be minimized.

The design issues can be coordinated as optimization issues. This issue and conditions are mentioned in Eqs. (2.2)–(2.5). To Minimize Area Control Error, based on

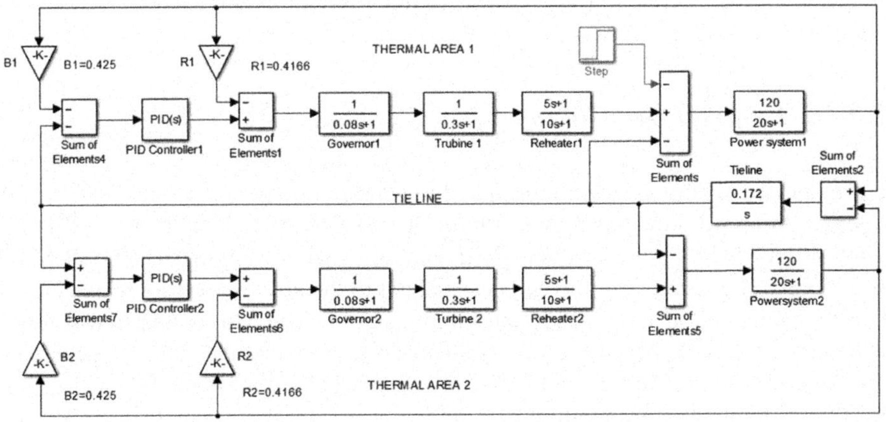

Fig. 2.1 Two areas interconnected thermal power system

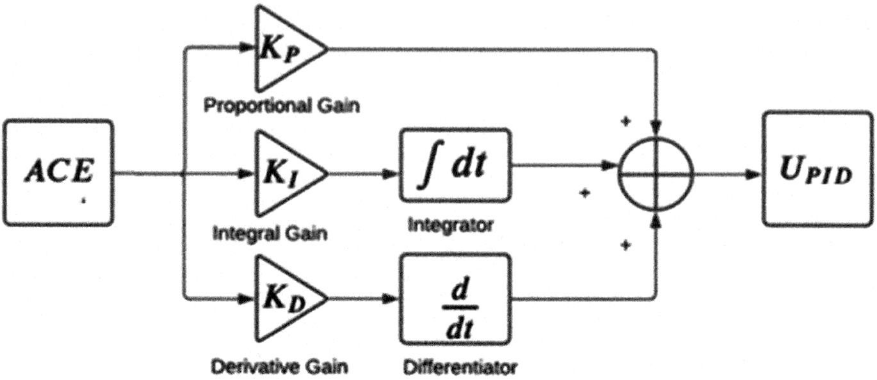

Fig. 2.2 General structure of PID controller

the limit of K_P min $\leq K_P \leq K_P$ max; K_I min $\leq K_I \leq K_I$ max and K_D min $\leq K_D \leq K_D$ max.

2.3 Swarm Intelligence Techniques

Swarm intelligence is a problem-solving approach inspired by the collective behavior of social insects and other animal groups. It involves the simulation of decentralized and self-organized systems where individual agents, such as particles or "swarm members," interact with one another and their environment to achieve a common goal. These algorithms use the principles of cooperation, communication, and adaptation to find optimal solutions to complex problems. Swarm intelligence techniques have applications in various fields, including optimization, robotics, blood glucose control in medical field, chemical process, energy, and artificial intelligence.

2.3.1 Artificial Bee Colony

Artificial Bee Colony Algorithm is one of the computing paradigms of swarm intelligence which was developed by Karaboga and Akay (2010), Naidu et al. (2014) in 2005. It has three varieties of bees in the colony such as employed, onlooker, and scout bees. This size of employed bees is equal to the size of food sources which is considered the initial solution. All the solutions get a chance to create a new solution in the first phase (employed bee phase) and a partner is randomly chosen to generate a new solution vector. Partner solution and the current solution should not be the same for the creation of a new solution in employed bees. Further, evaluate the objective function and fitness function of the newly generated solution. In all the swarm algorithms, the terms fitness and objective functions are equal but in ABC are not equal. In addition, the trial counter is used here to track the number of failures encountered by each solution. Once get a better solution the trial counter reset to zero. The employed bees are having one to one correspondence with the food source but onlooker bees are having correspondence with the food source depending on the probability. The probability values of all solutions are calculated before on looker phase. A solution with higher fitness values will have a higher probability. Also, the fitter solution may enter into the onlooker bee phase with multiple attempts. The employed bees memorize the food source message and convey positive feedback to onlooker bees (unemployed bees at the hive) in terms of the waggle dance. A scout, omitting the other's message, investigates everywhere in the hive randomly to generate new food source. Figure 2.3 depicts a flow chart illustrating the ABC (Artificial Bee Colony) algorithm (Naidu et al. 2014). This chart outlines the step-by-step process that the algorithm follows to solve optimization problems, mimicking the foraging behavior of bees in nature. The process of ABC algorithm flow is given below.

Step 1: Choose the swarm size (S = 5, 10, 15), number of iterations (T = 20), and limit = 1. Determine the size of employed bees (food source), i.e., Initial Population (Np = S/2).

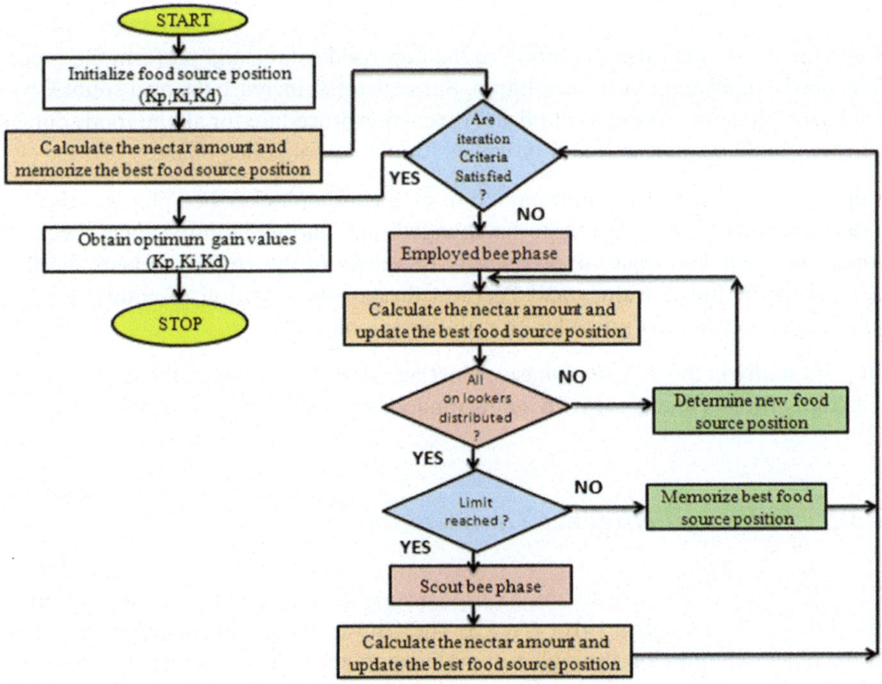

Fig. 2.3 Flowchart of ABC algorithm

Step 2: Create random solution vectors within the limit of decision variables and calculate their fitness values (f) of the initial population with the following expression as given below

$$\text{Fit} = \begin{cases} \frac{1}{1+f} & \text{if } f \geq 0 \\ 1 + |f| & \text{if } f < 0 \end{cases} \tag{2.6}$$

Step 3: Generate the initial trial vector for the population and select a random variable followed by a random partner.

Step 4: Create a new (solution) food vector in the employed bee phase using Eq. (2.7) and bound the solution if it is violated.

$$X^j_{new} = X^j + \emptyset\left(X^j - X^j_P\right) \tag{2.7}$$

where, X^j_{new}—jth variable of new solution

X^j —jth variable of current solution

X^j_P—jth variable of partner solution

$\emptyset$—Random values between -1 and 1.

Step 5: Evaluate the fitness solution of the new food source and perform the greedy selection to replace the current solution. Subsequently, increase the trial counter by 1 if it fails to generate a new solution and repeat this procedure for all the food sources. Further, calculate their probability values for the onlooker bee phase.

Step 6: Choose one of the current solutions for the onlooker bee phase and select a random number (r), variable to be changed, and partner solution then check the condition if r is less than the probability of values of the employed bees. Further, calculate their fitness values and Perform the greedy selection to update the best solution and repeat this procedure for all the onlooker bees.

Step 7: Finally in the scouting phase, select one solution for which the trial is greater than the limit and change it with a new random vector.

2.3.2 *Cohort Intelligence Optimization*

The CIO strategy is used for improving the quality of the power system by minimization of Area Control Error (ACE). The CIO technique enhances the behavior of the objective function through communication and tournament among its cohort (group of candidates) numbers. Figure 2.4 depicts the flowchart illustrating the CIO

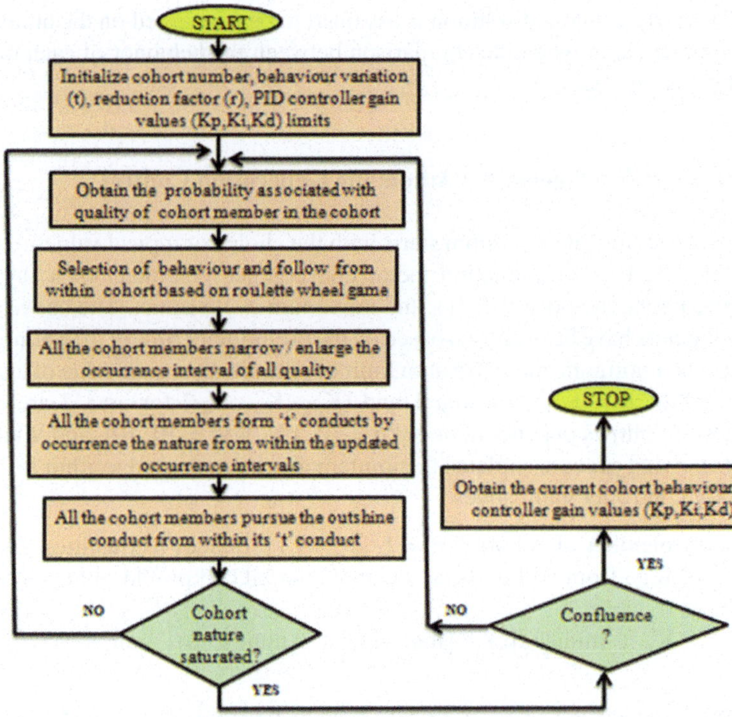

Fig. 2.4 Flowchart of CIO algorithm

algorithm (Murugesan et al. 2023a). The workflow of the CIO technique is addressed below.

Step 1: The auxiliary controller gain values are taken as virtue of candidates in the procedure of optimization paradigm. These virtues are randomly defined by the user in the search space between minimum and maximum allowable limits.

Step 2: The performance of the candidate is determined for all the group members, and is considered to be an objective (error) function. In the context of the automatic generation control, the objective function is the area control error operation to be diminished.

Step 3: The probability of every member of a cohort of being followed by another member of the cohort is calculated on the support of its objective function. The member of the cohort with the best objective function has the maximum chance of being followed by another member of the cohort.

Step 4: Each member of the cohort utilizes a roulette wheel selection to pursue a behavior in the cohort and it upgrades the solution of its objective function by increasing or decreasing the minimum and maximum limit of the auxiliary controller gain values.

Step 5: The newly generated solution is assumed to be converged on the attaining of the maximum cycle or when the comparison between the behavior of each member of the cohort is saturated.

2.3.2.1 Cohort Intelligence for Automatic Generation Control

The procedure of the CIO algorithm starts with the choice of typical values of cohort number (C = 5, 10, 15), reduction factor convergence parameter (r = 0.97) and convergence parameters ($\varepsilon = 0.001$), and maximum cycle Lmax. The above cohort design parameters have been chosen based on the preliminary runs of the optimization technique. The minimum and maximum limit of optimization gain value of auxiliary controllers (K_{p1}, K_{i1}, K_{d1}) for area 1 and (K_{p2}, K_{i2}, K_{d2}) for area 2 are chosen respectively as follows depending on $\Psi kp = [0, 10]$, $\Psi ki = [0, 10]$, and $\Psi kd = [0, 10]$. The following process explains the implementation cohort algorithm for tuning the auxiliary controller gain values.

Step 1: Every member of cohort c (c = 1, 2, … C) produces its qualities (K_{p1}, K_{i1}, K_{d1}, K_{p2}, K_{i2}, K_{d2}) from within the associated interval (Ψkp, Ψki, Ψkd) as below

$$K_P^C = \min(\Psi kp) + \left[\max(\Psi kp) - \min(\Psi kp)\right] \cdot \mathrm{ran}(.) \tag{2.8}$$

Step 2: The identical behavior of each member of cohort c (c = 1, 2, … C), i.e. Integral Time Absolute Error is obtained as follows:

$$\mathrm{ITAE}\ C = \int_0^t t \cdot |ACE_i(t)| \tag{2.9}$$

where i = 1 and 2 for two areas interconnected thermal power system

Step 3: Every member of the cohort determines the probability of its behavior as given below

$$PC = \frac{1/\mathrm{ITAE}^C}{\sum_{C=1}^{C} 1/\mathrm{ITAE}^C} \quad c = 1,\ 2,\ \ldots\ C \tag{2.10}$$

The probability of each cohort member aids in the pick a good solution.

Step 4: The roulette wheel method is used to select the best solution selection from current solutions. The solution with better performance is more chances of being followed by other members of the cohort.

Step 5: Each member in the cohort expands the sampling interval as follows

$$\psi kp \in \left[K_P^C - \left\|(\max(\psi kp) - \min(\psi kp))\frac{r}{2}\right\|, K_P^C + \left\|(\max(\psi kp) - \min(\psi kp))\frac{r}{2}\right\|\right] \quad (2.11)$$

Step 6: The result of saturation condition exists if there is no particular improvement in the member of the cohort as mentioned below

$$\left\|\max\left(ITAE^C\right)^n - \max\left(ITAE^C\right)^{n-1}\right\| \leq \varepsilon_1 \quad (2.12)$$

$$\left\|\min\left(ITAE^C\right)^n - \min\left(ITAE^C\right)^{n-1}\right\| \leq \varepsilon_2 \quad (2.13)$$

$$\left\|\max\left(ITAE^C\right)^n - \min\left(ITAE^C\right)^{n-1} \leq \varepsilon_3\right\| \quad (2.14)$$

Step 7: If either of the two choices mentioned below is true or false continue step 1.

1. A most number of attempt Lmax exceeded.
2. Cohort impregnate as related in Eqs. (2.12)–(2.14).

Choose any of the Cohort performances from the final optimized solution.

2.4 Simulation Results

The system model was crafted within the MATLAB Simulink environment, and Table 2.1 presents the finely-tuned regulator gain values. The model we constructed underwent simulation employing both the ABC and Cohort Intelligence optimization algorithms, subject to a one percent step load perturbation in area 1. The results, comprising the settling times for Automatic Generation Control comparisons in areas 1 and 2, as well as the tie-line data, can be found in Table 2.2.

Figures 2.5 and 2.6 illustrate bar chart analyses of the ABC and Cohort Intelligence optimization algorithms, respectively, applied to various cost functions within the proposed power system. In Fig. 2.5, it is evident that the ABC algorithm achieves the shortest settling time (under 30 s) in areas 1 and 2, as well as on the tie line when considering the ITAE function exclusively. Conversely, Fig. 2.6 demonstrates that the Cohort Intelligence algorithm yields improved system responses, with settling times of less than 30 s across all cost functions. Notably, the combination of CIO-ITAE and CIO-IAE in Fig. 2.6 results in the shortest settling time (less than 10 s) specifically in area 1.

Figures 2.7, 2.8 and 2.9 depict the assessment of the ABC algorithm's performance under various swarm sizes (5, 10, and 15) for area 1, area 2, and the tie line individually. These figures clearly illustrate that a swarm size of 15 leads to diminished oscillations on both the positive and negative sides and results in faster settling when compared to smaller swarm sizes. On the other hand, Figs. 2.10, 2.11 and

Table 2.1 Tuned gain values of ABC and cohort algorithm for various objective function

Algorithm	Gain values	ITAE	IAE	ITSE	ISE
ABC-PID	K_{P1}	4.5045	11.774	0.93773	25.4001
	K_{i1}	12.5388	3.4574	0.64762	4.343
	K_{d1}	0.65085	0.9409	0.99603	3.3719
	K_{p2}	1.776	5.9415	0.73622	2.7157
	K_{i2}	4.6549	3.381	0.05748	0.60257
	K_{d2}	2.4166	1.4809	0.00935	2.9015
CIO-PID	K_{p1}	9.3916	9.8577	8.2943	9.9522
	K_{i1}	9.8541	9.7857	8.2579	7.8564
	K_{d1}	1.8672	1.8169	7.7378	7.4484
	K_{p2}	6.6787	6.5218	9.2885	3.9312
	K_{i2}	0.3674	0.3911	4.0358	2.293
	K_{d2}	1.1061	5.4779	3.4539	8.6117

Table 2.2 Settling time of proposed system using ABC and cohort algorithm for various objective function

Algorithm	Response	Settling time in seconds			
		ITAE	IAE	ITSE	ISE
ABC-PID	Area 1 (Δf_1)	19	35	62	45
	Area 2 (Δf_2)	30	40	68	60
	Tie line (ΔP_{tie})	22	42	82	46
CIO-PID	Area 1 (Δf_1)	10	10	20	26
	Area 2 (Δf_2)	27	27	31	32
	Tie line (ΔP_{tie})	31	30	30	32

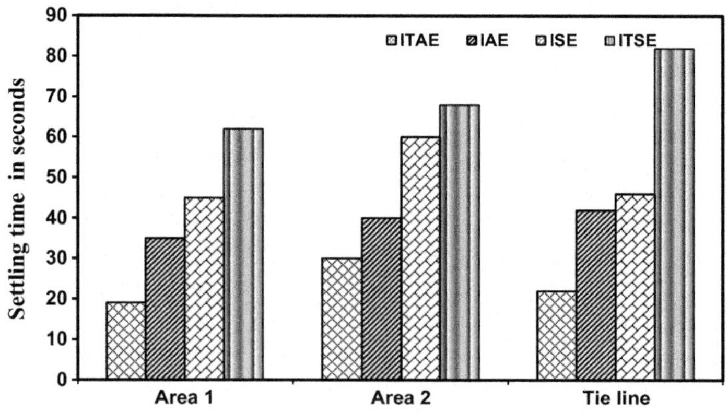

Fig. 2.5 Bar chart analysis of settling time for ABC algorithm

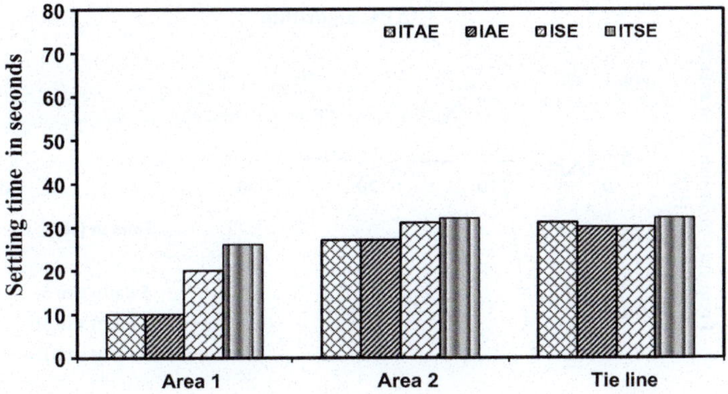

Fig. 2.6 Bar chart analysis of settling time for cohort algorithm

2.12 illustrate the evaluation of the cohort intelligence algorithm for different cohort numbers (5, 10, and 15) in area 1, area 2, and the tie line. These results highlight that variations in the cohort numbers during the optimization process do not yield significant improvements in the dynamics of the power system.

Figures 2.13, 2.14 and 2.15 provide a performance comparison between the ITAE-based PID controller optimized using both the ABC and the cohort optimization algorithm. These figures clearly demonstrate that the cohort intelligence approach is more effective in controlling oscillations in the dynamic simulation output when compared to the ABC-optimized controller.

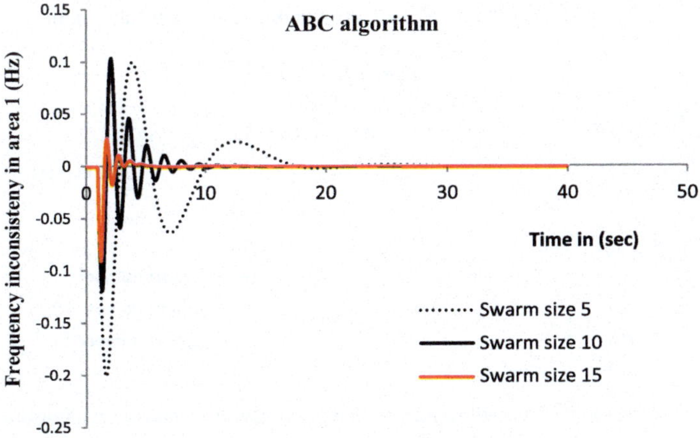

Fig. 2.7 Analysis of ITAE-based frequency changes in area 1 using ABC algorithm

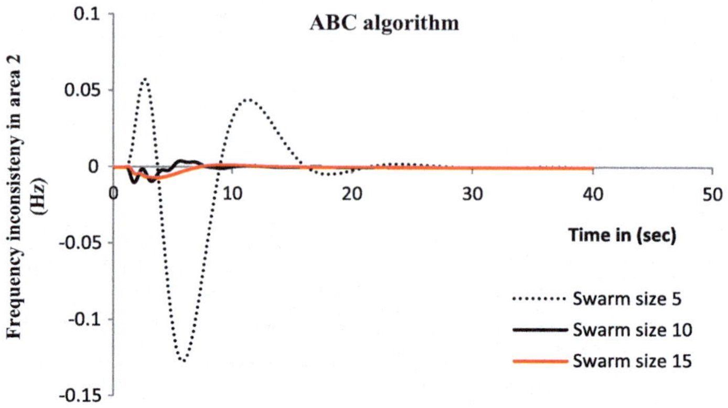

Fig. 2.8 Analysis of ITAE-based frequency changes in area 2 using ABC algorithm

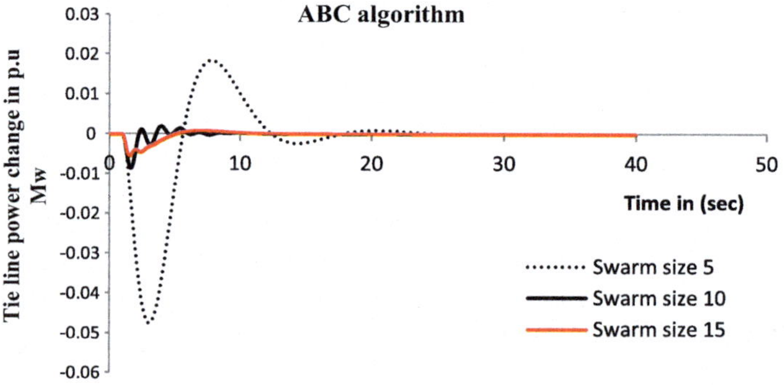

Fig. 2.9 Analysis of ITAE-based power changes in tie line using ABC algorithm

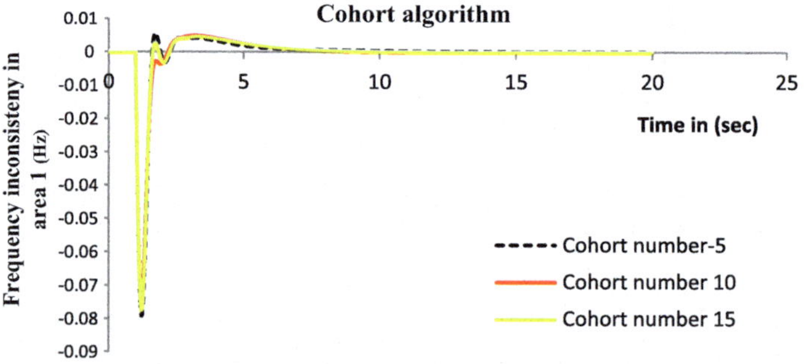

Fig. 2.10 Analysis of ITAE-based frequency changes in area 1 using cohort algorithm

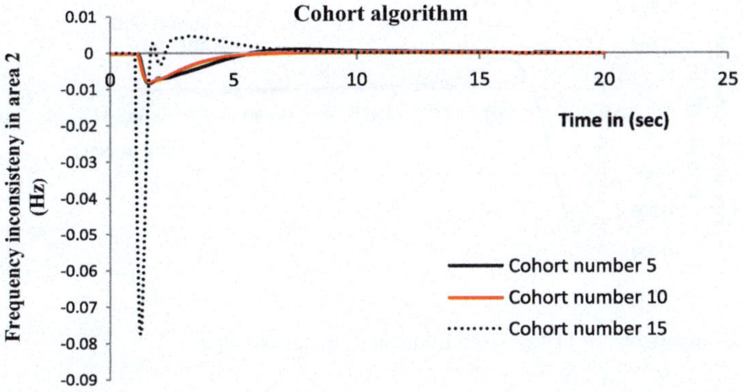

Fig. 2.11 Analysis of ITAE-based frequency changes in area 2 using cohort algorithm

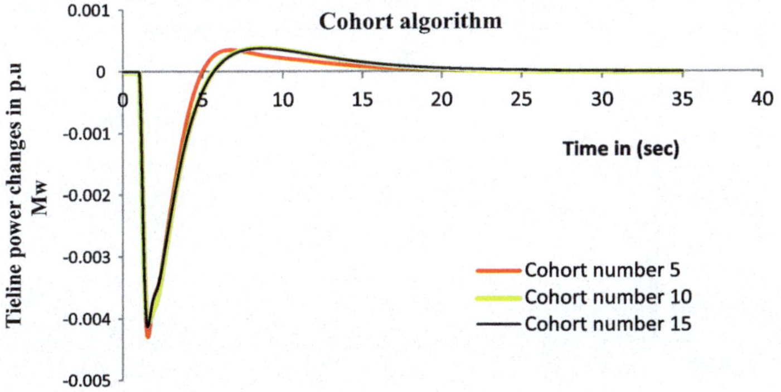

Fig. 2.12 Analysis of ITAE-based power changes in tie line using cohort algorithm

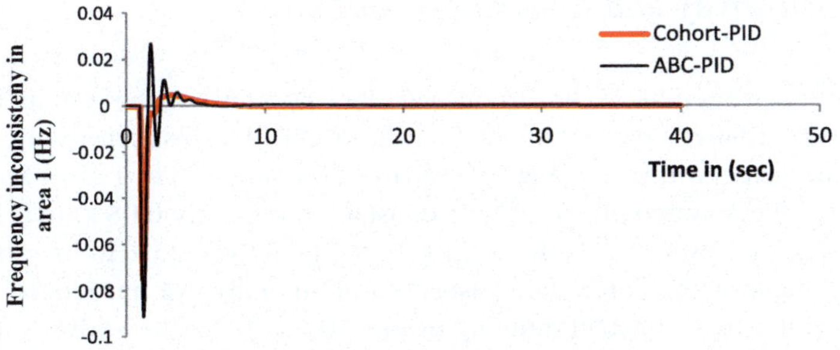

Fig. 2.13 Comparisons of ITAE-based frequency changes in area 1

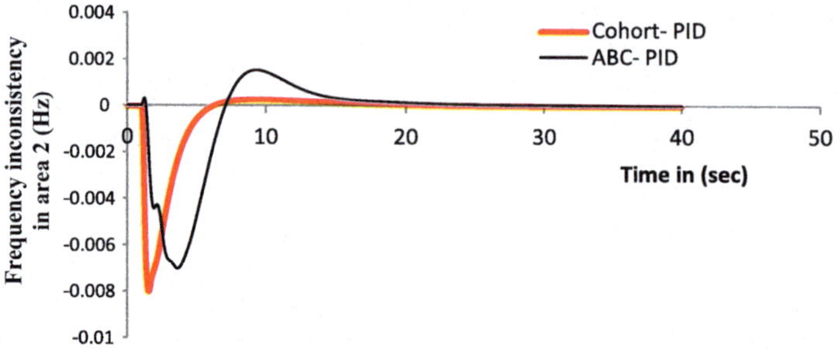

Fig. 2.14 Comparisons of ITAE-based frequency changes in area 2

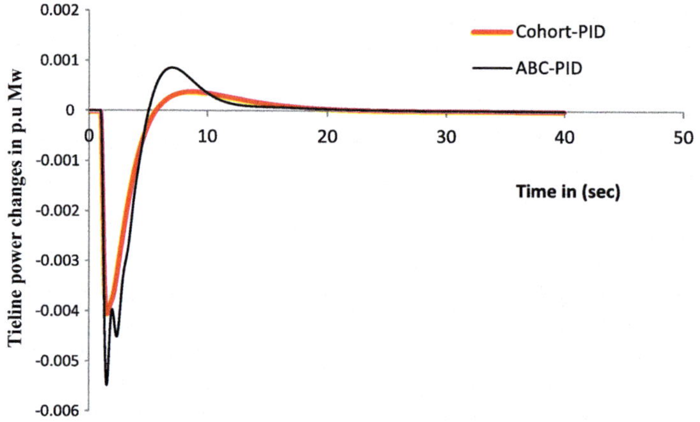

Fig. 2.15 Comparisons of ITAE-based frequency changes in tie line

2.4.1 Sensitivity and Robustness Analysis

In this section, we conducted sensitivity and robustness tests on the system framework using the ITAE-based cohort optimization algorithm. To assess sensitivity, we varied system parameters within the range of −50 to +50% and −25 to +25%. Specifically, we adjusted the nominal governor time constants (originally 0.08 s) to 0.12 s for + 50%, 0.04 s for −50%, 0.10 s for +25%, and 0.06 s for −25% to investigate their impact on the system's interrelated aspects. Additionally, we introduced variations in the generator time constant (ranging from −50% = 0.15 s to +50% = 0.45 s, and −25% = 0.225 s to +25% = 0.375 s) within the Simulink model to further assess system behavior. Furthermore, we modified the bias coefficients (ranging from − 50% = 0.2125 to −25% = 0.31875, with the nominal value at 0.425, and +25% = 0.53125 to +50% = 0.6375) to evaluate their impact on the system's performance.

Table 2.3 provides an overview of the time domain parameters, including undershoot (Ush) and settling time (Ts), illustrating variations between areas 1 and 2 and the tie line. This table demonstrates that the control technique based on the CIO algorithm has the capability to effectively handle sensitivity analysis. Figure 2.16 portrays the response to a random step load perturbation when utilizing the ITAE-based Cohort algorithm. This figure clearly illustrates that the proposed control methods not only excel in handling sensitivity but also exhibit robustness in effectively managing system disturbances.

Table 2.3 Sensitivity analysis of CIO algorithm based on ITAE cost function

System parameters	Sensitivity	Control Area 1		Control Area 2		Tie line	
		Ush	Ts	Ush	Ts	Ush	Ts
	Nominal (%)	−0.075	10	−0.008	28	−0.0045	31
Tg	+25	−0.08	9	−0.009	26	−0.0045	35
	−25	−0.08	9	−0.009	26	−0.0045	35
	+50	−0.0825	10	−0.010	28	−0.0050	35
	−50	−0.0625	12	−0.007	30	−0.0042	31
Tt	+25	−0.09	25	−0.011	32	−0.0055	34
	−25	−0.07	12	−0.007	27	−0.0042	31
	+50	−0.1	25	−0.015	25	−0.007	32
	−50	−0.045	10	−0.006	32	−0.0038	35
B	+25	−0.07	12	−0.005	34	−0.0035	60
	−25	−0.09	9	−0.013	28	−0.008	60
	+50	−0.06	12	−0.0035	30	−0.003	60
	−50	−0.11	18	−0.025	26	−0.008	28

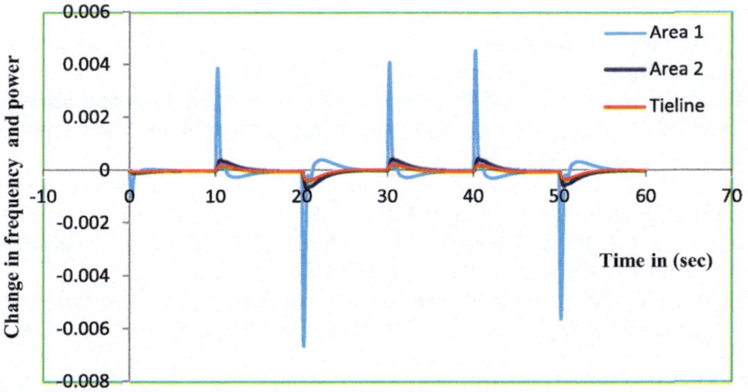

Fig. 2.16 Robustness analysis of CIO algorithm based on ITAE cost function

2.5 Conclusion

This paper introduces a PID auxiliary regulator for the Automatic Generation Control (AGC) of a two-area power system with a reheat turbine, where the areas are mutually dependent. The optimal PID regulator gain parameters were determined using two advanced techniques, employing four different error functions. The effectiveness of this intelligent optimization approach was demonstrated by analyzing time domain parameters, specifically undershoot and settling time. In this context, the performance of the CIO-ITAE combination outshone other objective functions and the ABC algorithm during sensitivity and robustness analysis. This analysis involved varying the time constants of the governor and turbine, adjusting the bias coefficient, and subjecting the system to random step load perturbations. The simulation results from both sensitivity and perturbation tests indicated that the CIO-PID regulators offered greater accuracy. Moreover, the CIO method exhibited insensitivity to wide-ranging variations in system parameters and demonstrated its ability to handle sudden changes. In the future, the effectiveness of cohort intelligence will be explored in the context of a hybrid controller integrated with a renewable energy system.

References

Aldeen M, Trinh H (1994) Load-frequency control of interconnected power systems via constrained feedback control schemes. Comput Electr Eng 20(1):71–88

Arora K, Kumar A, Kamboj VK (2019) Automatic generation control and load frequency control: a comprehensive review. In: Applications of computing, automation and wireless systems in electrical engineering: proceedings of MARC 2018, pp 449–456

Behera S, Sahoo S, Pati BB (2015) A review on optimization algorithms and application to wind energy integration to grid. Renew Sustain Energy Rev 48:214–227

Daneshfar F, Bevrani H (2012) Multiobjective design of load frequency control using genetic algorithms. Int J Electr Power Energy Syst 42(1):257–263

Debnath MK, Agrawal R, Tripathy SR, Choudhury S (2020) Artificial neural network tuned PID controller for LFC investigation including distributed generation. Int J Numer Model: Electron Netw, Devices Fields 33(5):e2740

Elgerd OI, Fosha CE (1970) Optimum megawatt-frequency control of multiarea electric energy systems. IEEE Trans Power Appar Syst 4:556–563

Jaleeli N, VanSlyck LS, Ewart DN, Fink LH, Hoffmann AG (1992) Understanding automatic generation control. IEEE Trans Power Syst 7(3):1106–1122

Jood P, Aggarwal SK, Chopra V (2019) Performance assessment of a neuro-fuzzy load frequency controller in the presence of system non-linearities and renewable penetration. Comput Electr Eng 74:362–378

Karaboga D, Akay BAHRİYE (2010) Proportional—integral—derivative controller design by using artificial bee colony, harmony search, and the bees algorithms. Proc Inst Mech Eng, Part I: J Syst Control Eng 224(7):869–883

Kulkarni AJ, Durugkar IP, Kumar M (2013, October) Cohort intelligence: a self supervised learning behavior. In: 2013 IEEE international conference on systems, man, and cybernetics. IEEE, pp 1396–1400

Kundur PS, Malik OP (2022) Power system stability and control. McGraw-Hill Education

Madasu SD, Kumar MS, Singh AK (2018) A flower pollination algorithm based automatic generation control of interconnected power system. Ain Shams Eng J 9(4):1215–1224

Mohanty B, Panda S, Hota PK (2014) Controller parameters tuning of differential evolution algorithm and its application to load frequency control of multi-source power system. Int J Electr Power Energy Syst 54:77–85

Moniya P, Anand B (2021) Automatic generation control with fuzzy logic controller incorporating tandem and cross compound turbine. J Ambient Intell Humaniz Comput 12:7071–7083

Murugesan D, Jagatheesan K, Shah P, Sekhar R (2023a) Fractional order PIλDμ controller for microgrid power system using cohort intelligence optimization. Results Control Optim 11:100218

Murugesan D, Jagatheesan K, Boopathi D (2021) Meta-heuristic strategy planned controller for frequency supervision of integrated thermal plant with renewable source. In: 2021 IEEE 3rd PhD colloquium on ethically driven innovation and technology for society (PhD EDITS), pp 1–2. https://doi.org/10.1109/PhDEDITS53295.2021.9649544

Murugesan D, Shah P, Jagatheesan K, Sekhar R, Kulkarni AJ (2022, September). Cohort intelligence optimization based controller design of isolated and interconnected thermal power system for automatic generation control. In: 2022 Second international conference on computer science, engineering and applications (ICCSEA). IEEE, pp 1–6

Murugesan D, Jagatheesan K, Kulkarni AJ, Shah P (2023). A socio inspired technique in nuclear power plant for load frequency control by using cohort intelligence optimization-based PID controller. In: Renewable energy optimization, planning and control: proceedings of ICRTE 2022. Springer Nature Singapore, Singapore, pp 1–12

Murugesan D, Jagatheesan K, Kulkarni AJ, Shah P, Sekhar R (2023) Comparison of biologically inspired algorithm with socio-inspired technique on load frequency control of multi-source single-area power system. In: Applied genetic algorithm and its variants: case studies and new developments. Springer Nature Singapore, Singapore, pp 185–208

Naidu K, Mokhlis H, Bakar AA (2014) Multiobjective optimization using weighted sum artificial bee colony algorithm for load frequency control. Int J Electr Power Energy Syst 55:657–667

Pappachen A, Fathima AP (2017) Critical research areas on load frequency control issues in a deregulated power system: a state-of-the-art-of-review. Renew Sustain Energy Rev 72:163–177

Pradhan PC, Sahu RK, Panda S (2016) Firefly algorithm optimized fuzzy PID controller for AGC of multi-area multi-source power systems with UPFC and SMES. Eng Sci Technol, Int J 19(1):338–354

Qian D, Zhao D, Yi J, Liu X (2013) Neural sliding-mode load frequency controller design of power systems. Neural Comput Appl 22:279–286

Shah P, Kulkarni AJ (2019) Application of variations of cohort intelligence in designing fractional PID controller for various systems. In: Socio-cultural inspired metaheuristics. Springer, Singapore, pp 175–192

Shankar R, Pradhan SR, Chatterjee K, Mandal R (2017) A comprehensive state of the art literature survey on LFC mechanism for power system. Renew Sustain Energy Rev 76:1185–1207

Chapter 3
Integration of Intelligent Systems for Efficient Smart Grid Management

O. Apata and Pitshou N. Bokoro

Abstract The integration of intelligent systems has become increasingly important in the management of the modern power grid, also known as the smart grid. The smart grid refers to the implementation of advanced technologies and communication systems that provide greater control, efficiency, reliability, and sustainability in the generation, transmission, distribution, and consumption of electricity. The integration of intelligent systems plays a crucial role in the effective management of the smart grid. One of the primary benefits of intelligent systems in the smart grid is improved efficiency, optimal energy consumption, and waste reduction. Additionally, intelligent systems can help detect and predict equipment failures, allowing for proactive maintenance and reducing downtime. Furthermore, intelligent systems can provide real-time monitoring of the smart grid, allowing for quick response to any issues that do arise. By analyzing energy consumption patterns, the smart grid can identify opportunities for energy conservation and reduce greenhouse gas emissions. The benefits of intelligent systems in the smart grid are numerous and far-reaching, providing a foundation for a more advanced and connected energy system of the future. Smart grids are designed to optimize the production, transmission, and consumption of electricity, thereby improving the reliability and efficiency of the power system. One of the key technologies that have been used to achieve this objective is Multi-Agent System (MAS). This chapter examines the role played by intelligent systems, especially multi-agent systems in the provision of extensible and flexible architecture for the deployment of intelligence in a smart grid.

Keywords Multi-agent systems · Renewable energy sources · Smart grid

O. Apata · P. N. Bokoro (✉)
Trilateral Research Chair in Transformative Innovation, The 4IR and Sustainable Development, College of Business and Economics, University of Johannesburg, Johannesburg 2028, South Africa

O. Apata
e-mail: g.apata@ieee.org

C. B. Khadse et al. (eds.), *Intelligent Methods in Electrical Power Systems*, Engineering Optimization: Methods and Applications, https://doi.org/10.1007/978-981-97-5718-3_3

3.1 Introduction

With the increasing demand for electricity, energy generation, transmission, and distribution have become complex processes, requiring advanced technologies and management techniques (Judge et al. 2022). Smart grids, a modernized and upgraded intelligent version of the traditional power grid, have emerged as a solution to meet the challenges of energy management. Smart grids use information and communication technologies (ICT) to monitor, control, manage the flow of electricity across the power grid, and have the capability to sense and measure power consumption (Yu and Hong 2015). Smart grids consist of various components such as smart meters, energy storage systems, inverters, and electric vehicles. These components generate and consume electricity, and their behavior affects the overall performance of the grid. Therefore, it is essential to manage the interactions among these components to ensure the efficient operation of the grid. The integration of intelligent systems in smart grids has become essential for efficient and sustainable energy management. The unique feature of the smart grid is applicable to various areas of energy production such as demand-side management, demand response, advanced metering, vehicle-to-grid capability, energy storage, renewable resources, and micro-grids (Deng et al. 2015). Smart grids offer several benefits over the traditional power grid. They use advanced technologies to monitor and control the flow of electricity, improving efficiency and reducing losses. Smart grids also enable the integration of renewable energy sources, reducing reliance on fossil fuels and promoting sustainable energy generation. Additionally, smart grids provide better visibility and control over energy consumption, enabling consumers to make informed decisions about their energy use. Smart grids can also improve the resiliency of the power grid, reducing the impact of power outages and other disruptions.

The role of intelligent systems and Artificial intelligence (AI) in smart grid systems is becoming increasingly important. Intelligent systems play a crucial role in smart grid management. They use advanced algorithms and data analytics to monitor and control the flow of electricity across the power grid. Intelligent systems can detect faults and anomalies in the power grid, enabling quick response and reducing downtime. They can also optimize energy generation and distribution, reducing losses and improving efficiency. AI plays a critical role in improving the efficiency and reliability of smart grid systems. Since smart grid systems are complex networks that rely on the seamless integration of various technologies, AI provide advanced analytics and automation capabilities that can help in the optimization of smart grids and improve system resilience. Some of the AI-based methods identified in various literatures include machine learning (ML) methods, expert systems (ES), artificial neural networks (ANN), fuzzy logic, and genetic algorithm (Bose 2017; Li et al. 2018; Xue et al. 2019). These techniques can be categorized as optimization-based, learning-based, rule-based, and agent-based. Overall, AI has the potential to transform smart grid systems by providing advanced analytics and automation capabilities that can improve efficiency, reliability, and resilience. However, to realize these benefits, there is a need to invest in the necessary infrastructure, data management, and personnel training to fully leverage these AI technologies.

The use of multi-agent systems (MASs) in smart grids has gained increasing attention in recent years due to the increasing complexity of modern power systems. The integration of renewable energy sources, distributed energy storage systems, and demand-side management into the power grid has made it necessary to develop new control and management strategies that can cope with the uncertainties and dynamics of these systems. MAS has been identified as a promising approach for addressing these challenges due to their ability to provide decentralized decision-making, coordination, and communication among the various agents in the system. MAS can help smart grids become more efficient, reliable, and sustainable by coordinating the actions of multiple agents, such as power generators, consumers, and energy storage devices, to achieve specific goals. They can play a critical role in the management and operation of smart grids by enabling coordination, optimization, flexibility, adaptability, security, and resilience. MAS is a promising technology that can help to achieve the various goals of smart grids such as flexibility and adaptability, security and resilience, optimization, coordination, and control.

3.1.1 Smart Grid

The traditional electricity grid has served us well for decades, but with the increasing complexities and challenges of the modern energy landscape, there is a need for a more advanced and intelligent electricity distribution system. This is where the concept of smart grids comes into play. A smart grid is an innovative and technology-driven electricity infrastructure that enhances the efficiency, reliability, and sustainability of the power grid. It integrates various advanced technologies, such as sensors, communication systems, control mechanisms, and data analytics, to optimize the generation, transmission, distribution, and consumption of electricity (Iwayemi et al. 2011). The smart grid is designed to improve the efficiency, reliability, and sustainability of the electricity system while accommodating the integration of renewable energy sources, electric vehicles, and other emerging technologies. The smart grid allows for two-way communication between utilities and consumers, enabling better management of electricity supply and demand (Colak et al. 2016). It also allows for real-time monitoring of the electricity grid, which can help to detect and respond to problems quickly, reducing the likelihood of power outages and improving system reliability. In addition, the smart grid can help to reduce carbon emissions by enabling the integration of renewable energy sources like solar and wind power into the electricity system. It also enables the use of electric vehicles as a grid resource, allowing them to be charged during off-peak hours when electricity demand is low, and providing grid services through vehicle-to-grid technology.

A smart grid comprises several interconnected components that work together to create an intelligent and dynamic energy ecosystem. The architecture of a smart grid includes three major components, namely, generation, transmission, and distribution as illustrated in Fig. 3.1 (Kabalci 2016). The generation component of a smart grid is responsible for the production of electricity from various sources, including traditional power plants and renewable energy installations. It focuses on optimizing the efficiency, reliability, and sustainability of power generation. The transmission component of the smart grid includes high-voltage power lines that transport electricity from power plants to distribution substations. A key aspect of the generation component in a smart grid is the integration of diverse generation sources. Traditionally, power grids heavily relied on centralized power plants, primarily based on fossil fuels such as coal, natural gas, and oil. However, a smart grid promotes the integration of renewable energy sources like solar, wind, hydro, geothermal, and biomass. This diversification of the generation mix reduces dependence on fossil fuels, decreases greenhouse gas emissions, and enhances the sustainability of the electricity supply. Smart grids facilitate the integration of distributed generation systems. Distributed generation involves the production of electricity closer to the point of consumption, reducing the need for long-distance transmission. It includes technologies such as rooftop solar panels, small wind turbines, micro-hydro installations, and combined heat and power (CHP) systems. Distributed generation enhances grid efficiency by minimizing transmission losses and optimizing energy use, while also improving the resilience and reliability of the power supply.

Advanced Metering Infrastructure (AMI) is a crucial component of smart grids (Le et al. 2016), particularly within the generation component. AMI refers to a system that replaces traditional electromechanical meters with advanced digital meters,

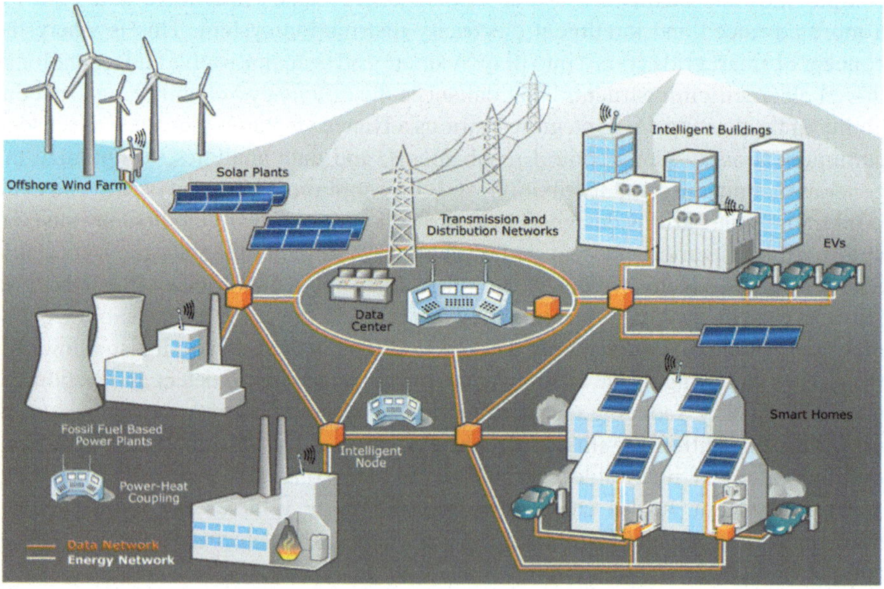

Fig. 3.1 Smart grid architecture (Kabalci 2016)

commonly known as smart meters. AMI enables two-way communication between utility providers and consumers, facilitating accurate and real-time monitoring of electricity consumption and supporting various grid management functions. AMI provides a range of functionalities beyond basic energy metering. Smart meters in AMI systems can measure and record electricity usage in intervals as short as 15 min, offering a much higher level of granularity compared to traditional meters. They transmit this consumption data back to the utility provider through secure communication networks. These meters enable utilities to implement time-of-use pricing (TOU), where electricity rates vary based on the time of day or season. Smart meters can record consumption data during different time periods, allowing utilities to charge higher rates during peak demand periods and lower rates during off-peak hours. TOU pricing incentivizes consumers to shift their electricity usage to off-peak periods, reducing strain on the grid during peak hours.

Energy storage plays a crucial role in the generation component of a smart grid. It addresses the intermittent nature of renewable energy sources and helps balance supply and demand. Smart grids enable the integration of various energy storage technologies, such as batteries, pumped hydro storage, flywheels, and compressed air energy storage (Tan et al. 2021). Energy storage systems store excess electricity during periods of low demand or high generation and release it during peak demand or low generation, ensuring a reliable power supply and grid stability.

Smart grids also incorporate advanced monitoring and control systems within the generation component. These systems provide real-time data on the performance, availability, and output of generation assets. Grid operators can monitor generation facilities, track renewable energy generation, and manage grid stability using accurate and timely information. This allows for optimized generation scheduling, efficient load balancing, and improved response to fluctuations in generation or changes in consumer demand.

The integration of advanced analytics and forecasting tools in smart grids enables better grid planning and optimization within the generation component. Grid operators can analyze historical data, consumption patterns, and generation outputs to forecast future demand accurately. This information helps in determining the optimal generation capacity, planning maintenance schedules, optimizing the dispatch of generation resources, and making informed decisions to ensure efficient and reliable electricity generation.

The transmission component is a vital part of the smart grid infrastructure that focuses on the high-voltage transmission of electricity from power plants to distribution networks (Rubino et al. 2021). It plays a crucial role in efficiently and reliably transporting electricity over long distances while maintaining grid stability. High-voltage transmission is essential to minimize power losses during long-distance electricity transport. The transmission component utilizes advanced analytics and grid planning tools to optimize the transmission infrastructure. Grid operators analyze historical data, consumption patterns, and generation outputs to forecast future demand accurately. They use this information to plan new transmission lines, identify potential congestion points, and optimize the routing of electricity. Grid planning and optimization ensure efficient utilization of transmission assets and minimize transmission losses. The interconnection of multiple power plants and renewable energy installations remains a very important part of the transmission process in a smart grid. It allows for the integration of electricity generated from different sources into the grid. The transmission system must be capable of handling the increased complexity and variability associated with the integration of renewable energy sources. This includes managing power fluctuations, optimizing power flow, and coordinating the dispatch of generation resources. The transmission component of a smart grid plays a critical role in integrating renewable energy sources. It facilitates the long-distance transmission of renewable electricity generated from remote areas to load centers. This integration requires careful planning and coordination to ensure grid stability, manage power fluctuations, and optimize power flow. Advanced forecasting and grid management tools are utilized to efficiently integrate and manage the variable output from renewable energy sources.

Smart grids incorporate wide-area monitoring systems within the transmission component. These systems utilize phasor measurement units (PMUs) and advanced sensors deployed at strategic locations across the transmission network (Song et al. 2017). PMUs measure voltage, current, and frequency at high sampling rates, enabling real-time monitoring of grid conditions over wide areas. Wide-area monitoring systems provide critical data for grid operators to assess grid stability, detect anomalies, and respond to contingencies. Smart grids also leverage advanced control and automation systems within the transmission component. These systems continuously monitor and control the flow of electricity to maintain grid stability and reliability. They employ technologies such as Supervisory Control and Data Acquisition (SCADA), remote terminal units (RTUs), and intelligent electronic devices (IEDs). Through automation, these systems can remotely operate transmission equipment, switch power lines, and control voltage and reactive power levels. Advanced protection and control systems within the transmission component is also crucial for wide-area protection and control. These systems employ sophisticated algorithms and communication networks to enable fast and coordinated responses to abnormal grid conditions or contingencies. Wide-area protection and control systems help mitigate the propagation of faults, ensure system stability, and minimize the risk of widespread blackouts.

As part of the transmission component, smart grids focus on modernizing and upgrading existing transmission infrastructure. This may involve retrofitting or replacing aging equipment, deploying advanced sensors and communication technologies, and adopting digital substation technologies. Grid modernization initiatives improve the efficiency, reliability, and resilience of the transmission system. By incorporating advanced monitoring, control, automation, and optimization techniques, the transmission component of a smart grid enables efficient and reliable high-voltage electricity transmission. It enhances grid stability, improves fault management, facilitates renewable energy integration, and supports the overall reliability and resilience of the electricity grid.

The distribution component of the smart grid infrastructure that focuses on delivering electricity from the transmission network to end-users, including residential, commercial, and industrial consumers. It ensures the efficient, reliable, and safe distribution of electricity while incorporating advanced technologies and strategies. Smart grids incorporate advanced distribution management systems (ADMS) within the distribution component (Mahela et al. 2020). ADMS integrates real-time data from various sources, including smart meters, sensors, and weather forecasts, to monitor and manage the distribution network. It provides grid operators with a comprehensive view of the distribution system, enabling efficient load balancing, fault detection, outage management, and restoration processes.

Smart grids employ distribution automation technologies within the distribution component to enhance grid reliability and efficiency. These technologies include advanced sensors, remote terminal units (RTUs), and intelligent electronic devices (IEDs). They enable real-time monitoring of grid conditions, fault detection, and rapid isolation of affected areas. Distribution automation systems can automatically reroute power, minimize outage durations, and improve service restoration during faults or outages.

The integration of distributed energy resources (DERs) is a key functionality of the distribution component of the smart grid. DERs include small-scale renewable energy systems, energy storage devices, and demand response resources. These resources generate or store electricity close to the point of consumption, reducing the need for extensive transmission infrastructure and improving grid efficiency. Smart grids facilitate the seamless integration and management of DERs, enabling efficient utilization of distributed generation, storage resources and focusing on enhancing grid resilience and incorporating microgrids. Microgrids are localized power systems that can operate independently or in conjunction with the main grid. They provide localized generation, storage, and distribution capabilities, ensuring a reliable power supply in specific areas or during emergencies. Microgrids within the distribution component enhance resilience, promote renewable energy integration, and enable load balancing at the local level.

Smart grids aim to modernize and upgrade the distribution infrastructure. This may involve replacing aging equipment, deploying advanced sensors and communication technologies, and adopting digital substation technologies. Grid modernization initiatives improve the efficiency, reliability, and resilience of the distribution system, ensuring smooth and reliable electricity delivery. By incorporating advanced monitoring, automation, fault management, and consumer engagement strategies, the distribution component of a smart grid enhances the efficiency, reliability, and resilience of electricity distribution. It enables efficient load balancing, improves outage management, promotes the integration of DERs, enhances consumer participation, and ensures a more sustainable and reliable electricity supply. Advanced sensors and meters measure parameters such as voltage, frequency, and harmonics. This enables utilities to monitor power quality, identify disturbances, and take corrective actions to maintain high-quality power supply. Power quality monitoring and management ensure reliable operation of sensitive equipment, reduce downtime, and improve overall consumer satisfaction.

Overall, smart grids are the future of the electricity distribution system, transforming the way we generate, transmit, and distribute electricity. By integrating advanced technologies, data analytics, and real-time monitoring and control systems, smart grids enhance the efficiency, reliability, and sustainability of the power grid. Smart grids offer numerous benefits, including enhanced grid reliability, reduced energy waste, improved response to outages, better integration of renewable energy, and increased consumer engagement. However, their successful implementation requires addressing challenges such as cost, data security, interoperability, consumer education, and regulatory frameworks. As we move towards a more sustainable and decentralized energy future, smart grids play a pivotal role in enabling efficient and reliable electricity distribution. With their ability to optimize generation, transmission, and distribution processes, smart grids pave the way for a greener, more resilient, and consumer-centric energy ecosystem. Embracing and investing in smart grid technologies will help us meet the growing energy demands, reduce carbon emissions, and create a more sustainable future for generations to come.

The concept of the smart grid encompasses various aspects, and is summarized as illustrated in Fig. 3.2 (Järventausta et al. 2010). It involves innovative approaches to infrastructure, particularly in future power distribution. This includes the utilization of power electronics and the incorporation of direct current (DC) technology.

By integrating active resources such as distributed generation, loads, energy storage systems, and electric vehicles, the traditional passive distribution network transforms into an active network. This shift necessitates the development of new network solutions and ICT (Information and Communication Technology) solutions for efficient network operation and asset management, providing intelligence to

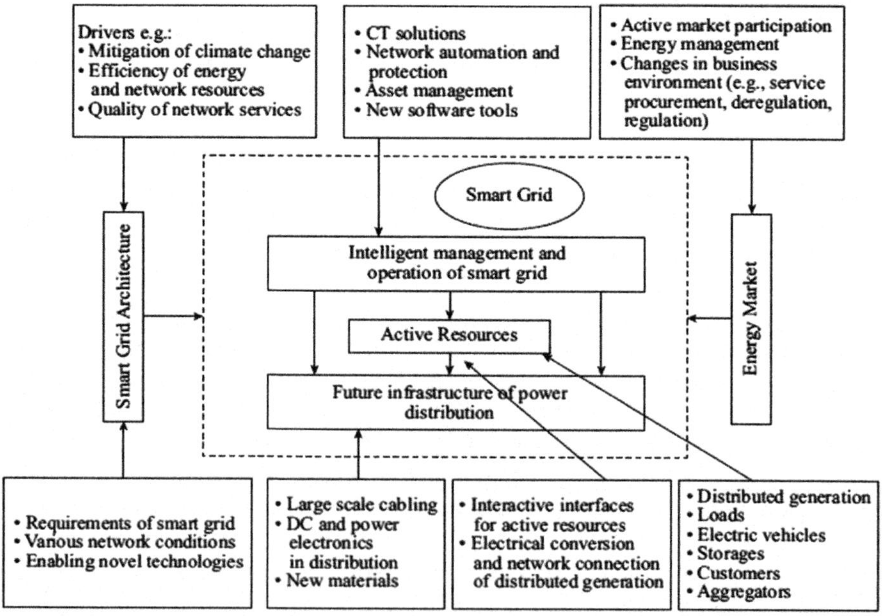

Fig. 3.2 Aspects of smart grid (Järventausta et al. 2010)

active networks. Smart grids facilitate the active participation of customers in the energy market and also influence changes in the business environment. They serve as customer-centric marketplaces for distributed generation and consumers of electricity. Figure 3.2 shows that the incorporation of distributed generation, electric vehicles, storage, predictive methods, demand response, and multidirectional power flow in a smart grid system enhances market integration, improves grid efficiency, and promotes the integration of renewable energy sources. These characteristics foster a more flexible, reliable, and sustainable electricity system.

To achieve effective control and management of the smart grid, a digital platform capable of handling communication protocols is necessary. This platform acts as an interface between the various components of the grid, including DGs, utility companies, and consumers. It facilitates the integration of existing transmission and distribution utilities into the smart grid, enabling enhanced flexibility, controllability, and data management (Li et al. 2010). The digital platform used for smart grid integration should be able to support various communication protocols to enable seamless data exchange between different devices, systems, and stakeholders in the grid. The platform should also be capable of handling large volumes of data generated by DGs, smart meters, and other grid components. It should provide data storage, processing, and analytics capabilities to extract meaningful insights for decision-making. Given the increased connectivity and digitalization in the smart grid, robust cybersecurity measures are crucial. The platform should incorporate strong security features to protect against potential cyber threats and ensure the integrity and privacy of the grid data.

Smart grids utilize advanced technologies and two-way communication to optimize energy distribution, integrate renewable sources, and enhance grid stability, while conventional grids lack these intelligent features and rely primarily on one-way energy flow (Ourahou et al. 2020). In conventional systems, market integration is typically weak, meaning there is limited coordination between electricity generators and consumers. However, in a smart grid system, market integration is improved by incorporating distributed power generation, which includes decentralized and renewable energy sources like solar panels and wind turbines. This allows for a more efficient and flexible electricity market. The smart grid incorporates both centralized and distributed generators. Centralized generators refer to large power plants that supply electricity to the grid, while distributed power generation refers to smaller-scale generation units located closer to the end consumers, such as residential solar panels or community wind farms. By incorporating distributed generation, the smart grid diversifies the energy sources and reduces reliance on centralized generation. The smart grid manages electricity generation by utilizing predictive methods to forecast demand accurately. This helps in optimizing the operation of power plants and distributed generators. Additionally, demand response programs enable consumers to adjust their electricity usage based on price signals or other incentives, allowing for more efficient use of resources and reducing peak demand.

Smart Grid Deployment in an existing power grid is a complex process that requires careful planning, design, and implementation. Here are some key steps that are typically involved in deploying a Smart Grid in an existing grid:

1. Conduct a Grid Assessment: A comprehensive assessment of the existing grid is required to determine its capability and limitations, identify potential issues and areas that need improvement, and evaluate the feasibility of a Smart Grid deployment.
2. Develop a Smart Grid Plan: Once the assessment is complete, a Smart Grid deployment plan should be developed. This plan should outline the key goals, objectives, and milestones of the project, as well as the specific technologies and infrastructure needed to support it.
3. Evaluate Technology Options: There are many different technologies and solutions available for Smart Grid deployment, including advanced metering infrastructure, demand response systems, distribution automation, and energy storage. The appropriate technology options should be evaluated and selected based on the specific needs and requirements of the grid.
4. Design and Implement Smart Grid Infrastructure: Once the technology options are selected, the next step is to design and implement the Smart Grid infrastructure. This includes deploying new sensors, communication networks, and control systems to monitor and manage grid operations.
5. Test and Validate Smart Grid Functionality: Before going live, it is important to thoroughly test and validate the Smart Grid's functionality. This includes testing individual components, as well as the integrated system.
6. Go Live with the Smart Grid: After successful testing and validation, the Smart Grid can be deployed live on the existing grid. This includes the deployment of new meters, communication networks, and control systems, as well as training for utility staff and customers on how to use the new system.
7. Monitor and Maintain Smart Grid Performance: Once the Smart Grid is live, ongoing monitoring and maintenance is required to ensure optimal performance. This includes monitoring grid performance, analyzing data, and identifying areas for improvement or optimization.

Overall, the deployment of a smart grid in an existing power grid requires careful planning, selection of appropriate technologies, and careful implementation and ongoing monitoring to ensure optimal performance. Smart grids are characterized by several features, including:

1. Integration of renewable energy sources: Smart grids are designed to integrate renewable energy sources such as solar and wind power into the grid system. This helps to reduce the reliance on fossil fuels and minimize the carbon footprint of the power sector.
2. Two-way communication: Smart grids facilitate two-way communication between the grid operators and the end-users. This enables real-time monitoring and control of the power system, allowing for more efficient use of energy resources.
3. Advanced metering: Smart grids use advanced metering systems that allow for the measurement and recording of electricity usage in real-time. This helps to promote energy conservation and reduce wastage.

4. Intelligent distribution: Smart grids use intelligent distribution systems that enable the efficient and reliable distribution of electricity to end-users. This helps to minimize power outages and improve the overall reliability of the power system.
5. Demand response: Smart grids enable demand response programs that encourage end-users to reduce their energy consumption during peak periods. This helps to reduce the strain on the power system and promote energy conservation.

3.1.2 Multi-agent Systems

A multi-agent system (MAS) is a collection of multiple autonomous agents, each with its own individual goals and objectives, which interact with each other to achieve common or individual goals. A multi-agent system is a type of distributed system in which the agents work together to achieve (Amirkhani and Barshooi 2022) a common objective or set of objectives, without any central control. The agents in a MAS can be software agents, robots, humans, or a combination of these. The concept of multi-agent systems is derived from the field of artificial intelligence and cognitive science. MAS can be found in various domains and applications, including robotics, artificial intelligence, economics, social sciences, transportation systems, and more. In the early days of AI research, the focus was on building intelligent systems that could perform a wide range of tasks. However, it soon became clear that building an intelligent system that could perform all tasks was not possible. Instead, it was more effective to build a system of intelligent agents that could work together to solve a problem.

The key idea behind MAS is to study and design systems where multiple agents can work together, collaborate, or compete to solve complex problems that may be difficult or even impossible for a single agent to accomplish alone. By leveraging the collective intelligence and abilities of multiple agents, MAS enables the system to exhibit emergent behavior and achieve outcomes that surpass what a single agent could achieve. One of the key advantages of multi-agent systems (MASs) is their inherent flexibility and adaptability. The autonomy of individual agents allows them to adjust their behavior and strategies in response to changes in the environment, the behavior of other agents, or even their own internal states. The flexibility and adaptability of MAS make them suitable for dynamic and complex domains, where the ability to respond and adapt to changing conditions is critical. By leveraging the autonomous nature of individual agents, MAS can exhibit robustness, scalability, specialization, learning, and effective coordination, ultimately leading to improved performance and problem-solving capabilities.

In a multi-agent system, the agents communicate with each other and exchange information to achieve their goals (Albagli et al. 2016). The communication can be either direct or indirect, and can be done through a variety of means, including messaging, email, shared databases, and so on. The choice of communication mechanism depends on the nature of the MAS, the capabilities of the agents, and the context in which they operate. Different mechanisms can be combined, and the communication strategies can evolve dynamically based on the changing requirements and conditions of the MAS. It's important to note that communication in MAS can involve not only the exchange of explicit information but also the sharing of beliefs, goals, intentions, and social norms, which are essential for effective collaboration and coordination among the agents. In some cases, agents can learn to communicate with each other by developing their own communication protocols. Through machine learning techniques or evolutionary algorithms, agents can discover effective ways to exchange information and coordinate their behaviors over time. This allows for the emergence of communication systems that are optimized for the specific tasks or goals of the MAS.

The agents in a MAS can also be organized into hierarchies or teams, which can work together to achieve common goals. Organizing agents into hierarchies or teams is a common approach in multi-agent systems to facilitate collaboration and achieve common goals more effectively. Hierarchies are beneficial as they allow for efficient coordination, information flow, and control within the MAS. In a hierarchical MAS, agents are organized into a structure resembling a hierarchy, where agents are grouped into levels based on their roles, responsibilities, or expertise. The hierarchy can have multiple levels, with higher-level agents supervising and coordinating the actions of lower-level agents. Each level of the hierarchy may have its own specific goals, and agents at different levels have different decision-making authority. Higher-level agents can provide guidance, allocate tasks, and aggregate information from lower-level agents, while lower-level agents can focus on executing their assigned tasks. Hierarchies provide a clear division of labor, reduce the complexity of decision-making, and enable scalability in large-scale MAS.

In a team-based MAS, agents are grouped into teams or coalitions, where each team consists of agents that share a common goal or have complementary capabilities. Teams can be formed based on various criteria such as expertise, task similarity, or geographical location. Agents within a team collaborate closely, communicate effectively, and work together to achieve their shared objectives. Teams promote specialization and cooperation within the MAS. Agents within a team can divide the workload, share information, coordinate their actions, and provide mutual support. By leveraging the diverse skills and knowledge of team members, teams can tackle complex tasks more efficiently and effectively.

Hierarchies and teams can be combined within a MAS, with hierarchies providing overall coordination and teams focusing on specific tasks or problem domains. This combination allows for flexible and adaptive organization, where hierarchies provide high-level control and coordination, while teams enable focused collaboration and specialization. It's worth noting that the specific organizational structure within a MAS can vary depending on the application and domain. The choice between hierarchies, teams, or a combination thereof depends on factors such as task complexity, scalability requirements, communication overhead, and the degree of autonomy desired for the individual agents.

By leveraging agent autonomy, communication, and coordination, MAS enables fault detection and diagnosis, efficient management of distributed energy resources, effective demand response, robust cybersecurity, and improved voltage regulation. These advancements contribute to the development of more reliable, secure, and sustainable energy systems. MAS can assist in the detection and diagnosis of faults within the smart grid infrastructure. Agents can monitor sensors, meters, and devices across the grid, analyzing data and identifying anomalies or deviations from expected behavior. By detecting faults in real-time, MAS enables prompt response, maintenance, and restoration of the grid, reducing downtime and improving overall reliability.

The management and coordination of distributed energy resources (DERs) in smart grids is facilitated by MAS. Agents representing DERs, such as solar panels, wind turbines, or energy storage systems, can communicate and collaborate to optimize energy generation, storage, and consumption. MAS-based distributed energy management maximizes the utilization of renewable energy sources, minimizes grid congestion, and improves overall energy efficiency.

Effective demand response management in smart grids is also enabled by MAS. Agents representing consumers, energy providers, and grid operators can communicate and negotiate to adjust energy consumption in response to price signals or grid conditions. MAS-based demand response management optimizes load balancing, reduces peak demand, and enhances grid stability while allowing consumers to actively participate in managing their energy usage.

MAS plays a crucial role in ensuring the cybersecurity of smart grid systems. Agents can be deployed to monitor network traffic, detect anomalies, and prevent or respond to cyber threats and attacks. By employing MAS for cybersecurity, smart grids can enhance their resilience against hacking, unauthorized access, data breaches, and other cyber threats, safeguarding the integrity and reliability of the grid infrastructure.

MAS can aid in voltage regulation within smart grids. Agents representing voltage regulators, capacitors, or distributed generators can coordinate their actions to maintain voltage levels within the desired range. MAS-based voltage regulation optimizes power quality, minimizes voltage fluctuations, and improves the overall stability and reliability of the grid.

3.1.3 Architecture of Multi-agent Systems

Multi-Agent Systems (MASs) can be broadly classified into three main categories based on their architectural designs namely, centralized, distributed, and hierarchical architecture (Mahela et al. 2020). It's worth noting that these architectural classifications are not mutually exclusive, and many real-world systems can exhibit characteristics of more than one category. The choice of MAS architecture depends on the specific requirements, complexity, and dynamics of the environment in which the agents operate.

The architecture of a multi-agent system consists of three main components: agents, environment, and communication. The behavior of the agents in a MAS is very important. These are autonomous entities that can perceive their environment, make decisions, and take actions to achieve their goals. They are the fundamental building blocks of a multi-agent system. Each agent may have its own internal state, knowledge, and reasoning abilities. The environment on the other hand is the external world in which the agents operate, and can include physical objects, other agents, and virtual spaces, providing the agents with information through perception, and while agents interact with the environment by performing actions, the environment can be dynamic, stochastic, and subject to change based on the actions of agents. The communication in a MAS is the means by which the agents interact with each other, and can include direct communication, such as messaging and speech, as well as indirect communication, such as shared databases and blackboards.

These three components work together to enable the functioning of a multi-agent system. Agents perceive the environment, make decisions based on their internal state and knowledge, and interact with the environment through actions. They also communicate and collaborate with other agents to achieve common objectives or resolve conflicts. The architecture of a multi-agent system provides a framework for organizing these components and designing the interactions and behaviors of the agents to achieve desired outcomes. Though the environment and communication are important components of a MAS, this chapter will focus on the agents in the MAS. There are different types of agents in a MAS, including reactive agents, deliberative agents, and hybrid agents (Wooldridge 2009). Each of these will be discussed briefly in the following subsections.

3.1.3.1 Reactive Agents

In a MAS, reactive agents are agents that are designed to respond to changes in their environment in real-time. These agents typically have a set of pre-defined rules or behaviors that determine how they should react to certain stimuli. Reactive agents are useful in multi-agent systems because they can quickly adapt to changes in the environment without needing to reason about complex models or make long-term plans. Instead, they rely on simple rules and heuristics to make decisions in real-time. However, reactive agents have some limitations. They are often unable to coordinate effectively with other agents in the system and may not be able to handle complex or unexpected situations. To overcome these limitations, some multi-agent systems use a combination of reactive and deliberative agents, where deliberative agents are responsible for higher-level planning and coordination.

3.1.3.2 Deliberative Agents

Deliberative agents are more complex and are capable of reasoning and planning to achieve their goals. Hybrid agents combine the reactive and deliberative approaches and can switch between them depending on the situation. In a MAS, deliberative agents are agents that are designed to reason and make decisions based on their current beliefs, goals, and plans. These agents typically have a more complex decision-making process than reactive agents and are better equipped to handle unexpected situations. Deliberative agents typically have a model of the world that they use to reason about their environment and plan their actions. They can use this model to generate multiple possible courses of action, evaluate them based on their goals, and choose the best one. Deliberative agents are useful in multi-agent systems because they can handle more complex and dynamic environments than reactive agents. They can reason about their actions and coordinate with other agents in the system to achieve shared goals. However, deliberative agents may also be slower to respond to changes in the environment and require more computational resources than reactive agents.

3.1.3.3 Hybrid Agents

Hybrid agents are useful in situations where the environment is both dynamic and complex. By combining reactive and deliberative behaviors, they can quickly respond to changes in the environment while still making use of more sophisticated reasoning when needed. Hybrid agents can provide a more flexible and adaptable approach to multi-agent systems than relying solely on reactive or deliberative agents. They can make use of both approaches to handle different aspects of the environment and achieve their goals more effectively. However, designing and implementing hybrid agents can be more complex than using either reactive or deliberative agents alone. The agent's behaviors and decision-making processes must be carefully designed and balanced to ensure that they work together effectively.

3.1.4 Multi Agents in Smart Grids

The performance of any power system is largely dependent on its control scheme which is usually a combination of hardware protocols for signal exchange or control and appropriate software. In modern power systems, such controls are carried out by a supervisory control and data acquisition (SCADA) system (Manickavasagam 2014). SCADA systems have the potential to be victim of cyber-attacks therefore raising questions around grid reliability. For smart grids, there is the need for advanced and fast sensing, control, and communication technologies. The MAS fits into this requirement and is therefore a veritable tool for sophisticated supervisory and control of the smart grid, making it easy to solve complex problems intelligently and efficiently.

3.1.4.1 Smart Grids and MAS

Smart grids and multi-agent systems (MASs) are closely intertwined and can greatly benefit from each other's capabilities. Smart grids leverage MAS to achieve efficient and autonomous operation, while MAS provides the necessary coordination and decision-making mechanisms within smart grid environments. In a smart grid context, intelligent agents can be classified into several categories based on their functions and capabilities. Here are some of the most common classifications:

1. Energy Management Agents (EMA): These agents are responsible for managing the energy consumption and production of the smart grid system. They can monitor the energy demand and supply in real-time and take actions to balance them by adjusting the production and consumption of energy.
2. Communication Agents: These agents facilitate communication between various components of the smart grid system, such as sensors, devices, and other agents. They ensure that the information is transmitted accurately and in a timely manner, which is critical for the smooth functioning of the system.
3. Grid Stability Agents: These agents are responsible for maintaining the stability and reliability of the smart grid system. They monitor the system for any anomalies or disturbances and take corrective actions to prevent or mitigate any potential problems.
4. Market Agents: These agents are responsible for managing the energy market by matching the energy supply and demand and ensuring that the energy is traded efficiently and fairly.
5. Security Agents: These agents are responsible for ensuring the security of the smart grid system by detecting and preventing cyber-attacks and other security threats.
6. Customer Agents: These agents interact directly with customers to provide them with information on their energy usage and help them manage their energy consumption more efficiently.

These intelligent agents listed above can be broadly classified into two categories. One relating to energy pricing, energy planning, energy market, and management of energy balance sheets while the other focuses on grid efficiency, security, fault handling, and dependability. Market agents focus on activities related to energy markets, pricing, and economic aspects of the smart grid. They represent various market participants, such as energy producers, consumers, aggregators, and brokers. Market agents are involved in bidding, negotiation, and trading activities to buy or sell energy based on market conditions, supply and demand dynamics, and pricing mechanisms. They facilitate efficient energy markets, optimize resource allocation, and ensure economic viability within the smart grid ecosystem. Grid management agents are responsible for the operational aspects of the smart grid. They focus on grid efficiency, security, fault handling, and overall system management. These agents monitor grid conditions, analyze data from sensors and smart devices, and take actions to optimize grid performance, maintain stability, and handle faults or disruptions. Grid management agents are involved in tasks such as load balancing, voltage regulation, fault detection and management, and coordination of grid resources.

While this classification highlights two broad categories, it's important to note that there can be overlap and interaction between the functions of these agent categories. For example, grid efficiency and security agents may also participate in energy markets to optimize their grid operations, or energy market agents may consider grid constraints and stability while optimizing energy trading. The ultimate goal of both categories of intelligent agents is to ensure the efficient, reliable, and secure operation of the smart grid. By focusing on different aspects of the grid's functioning, these agents contribute to the overall optimization and performance of the smart grid ecosystem.

3.1.4.2 MAS and Control of Smart Grids

To achieve a coordinated control of a smart grid, there is a need for the agents in the smart grid to work in a hierarchical way in controlled layers. Agent-based control of the smart grid is achieved by a seven-layered communication architecture as shown in Fig. 3.3. Each of these layers is described briefly below.

In a smart grid, the main grid agents are the central controllers responsible for managing the flow of electricity across the entire grid. These agents use sophisticated algorithms and real-time data to make decisions about how to optimize the distribution of electricity to ensure that the system operates efficiently, reliably, and securely.

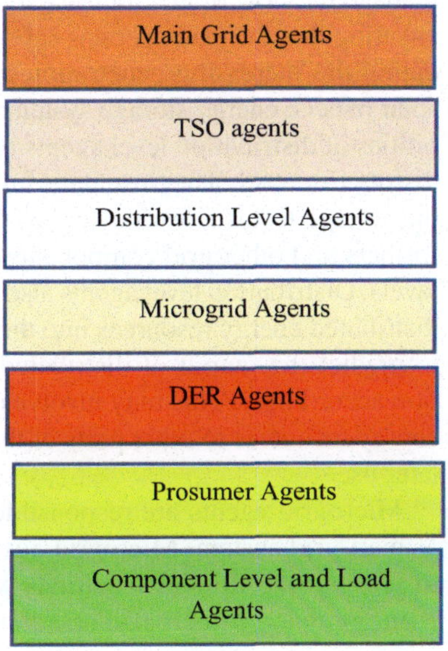

Fig. 3.3 Seven layered agent-based control structure of smart grid

The main grid agents communicate with individual grid components, such as power plants, substations, and smart meters, to gather information about the electricity supply and demand at each location. They then use this data to make decisions about how much electricity to generate, where to direct it, and when to do so. In addition to their role in managing the flow of electricity, main grid agents can also play a crucial role in optimizing the use of renewable energy sources, such as wind and solar power. By using data analytics and advanced forecasting techniques, these agents predict the availability of renewable energy sources and adjust the distribution of electricity accordingly.

The transmission system operator (TSO) agents are responsible for managing the transmission grid in a smart grid system. The transmission grid is the high-voltage network that transports electricity from power generation facilities to distribution networks, where it is then delivered to homes and businesses. TSO-level agents use sophisticated algorithms and real-time data to manage the flow of electricity across the transmission grid. They are responsible for ensuring that the grid operates within safe operating limits, maintaining the reliability and stability of the system. One of the key responsibilities of TSO-level agents is to balance the supply and demand of electricity in real-time. They monitor the grid's electricity flow, identifying any imbalances between supply and demand and taking corrective action, as necessary. This may involve adjusting the output of power generation facilities, directing electricity from one part of the grid to another, or asking consumers to reduce their energy consumption.

Distribution level agents are responsible for managing the distribution grid in a smart grid system. The distribution grid is the lower-voltage network that delivers electricity from the transmission grid to individual consumers, such as homes, businesses, and factories. These agents use real-time data and advanced algorithms to manage the flow of electricity across the distribution grid. They are responsible for ensuring that the grid operates safely, reliably, and efficiently, while also accommodating the increasing penetration of distributed energy resources, such as rooftop solar panels, energy storage systems, and electric vehicles. One of the key responsibilities of distribution level agents is to monitor and manage the voltage levels of the distribution grid. They use data from smart meters and other sensors to determine the voltage levels at various points on the grid and adjust the output of power transformers and other grid components as necessary to maintain safe and stable voltage levels. Distribution level agents also play a critical role in managing the integration of distributed energy resources into the grid. They use advanced forecasting techniques to predict the output of distributed energy resources and adjust the grid's operation accordingly. This may involve directing excess energy generated by distributed energy resources to other parts of the grid or storing it in energy storage systems for later use.

Microgrid agents are responsible for managing the operation of a microgrid in a smart grid system. Microgrid agents use real-time data and advanced algorithms to manage the flow of electricity within the microgrid. They are responsible for ensuring that the microgrid operates efficiently, reliably, and sustainably, while also supporting the integration of distributed energy resources. One of the key responsibilities of microgrid agents is to balance the supply and demand of electricity within the microgrid. They use data from smart meters and other sensors to determine the electricity consumption and generation within the microgrid and adjust the output of distributed energy resources and energy storage systems as necessary to maintain a balance between supply and demand.

Distributed Energy Resource (DER) agents are responsible for managing the operation of distributed energy resources in a smart grid system. Distributed energy resources are small-scale energy sources, such as solar panels, wind turbines, and energy storage systems, that are located close to the point of consumption, rather than at a centralized power plant. DER agents use real-time data and advanced algorithms to manage the operation of distributed energy resources. They are responsible for ensuring that the distributed energy resources operate efficiently, reliably, and sustainably, while also supporting the integration of renewable energy sources into the grid.

Prosumers are consumers who also produce energy, typically using distributed energy resources, such as solar panels and wind turbines. Prosumer agents also play a critical role in managing the integration of prosumers into the grid. They use advanced forecasting techniques to predict the energy production and consumption of prosumers and adjust the operation of the distributed energy resources accordingly. This may involve exporting excess energy generated by prosumers to the main grid or storing it in energy storage systems for later use.

Component level and load agents also play a critical role in maintaining grid stability and reliability. They use advanced analytics and predictive modeling to identify potential issues before they occur and take proactive measures to prevent these issues from impacting the grid. One of the key responsibilities of component level and load agents is to balance the supply and demand of electricity in the grid. Component level agents are responsible for managing the operation of specific components, such as transformers, capacitors, and switches, in the grid. Load agents, on the other hand, are responsible for managing the energy consumption of specific loads in the grid. They use real-time data from smart meters and other sensors to monitor the energy consumption of individual loads and adjust the operation of these loads as necessary to reduce energy consumption and support grid stability.

3.1.5 Implementation of MAS in Smart Grids

The implementation of MAS in smart grids involves the following steps.

Agent Design: The first step in implementing MAS in smart grids is to design the agents. Agents are the software entities that represent the various devices and stakeholders in the grid. Agents can be designed to represent devices such as smart meters, inverters, and energy storage systems. Agents can also represent stakeholders such as consumers, producers, and grid operators.

Agent Communication: The next step is to enable communication among the agents. Agents can communicate with each other using a communication protocol such as the Message Passing Interface (MPI) or the Simple Object Access Protocol (SOAP). Communication among the agents enables them to share information and coordinate their actions.

Agent Cooperation: The third step is to enable cooperation among the agents. Cooperation among the agents involves the sharing of information and the coordination of actions to achieve a common goal. For example, agents can cooperate to manage the power flow in the grid by balancing the supply and demand of electricity.

Agent Learning: The fourth step is to enable learning among the agents. Learning among the agents involves the adaptation of behavior based on the feedback received from the environment. For example, agents can learn to adjust their energy usage based on the price of electricity.

In future power systems, the utilization of smart grid technologies, such as automatic voltage regulators, smart meters, and sensors, will significantly improve the grid's capabilities. These enhancements include grid optimization, self-recovery, dynamic pricing, seamless integration of renewable energy sources, and efficient utilization of storage devices (Sharma et al. 2016). Upgrading the communication and information systems enables the control of these advanced grids, and multi-agent systems (MASs) play a crucial role in achieving coordinated control of smart grids.

3.1.5.1 Performance Improvement in Smart Grid Using MAS

There are several ways in which MAS can be used to improve the performance of smart grids. These agents can communicate with each other and make decisions based on local information and global objectives, leading to improved efficiency, reliability, and cost-effectiveness of the power system. Here are some potential ways a smart grid can be improved using MAS:

1. Optimal resource allocation: In a smart grid, there are various resources such as energy generation units, energy storage units, and loads. MAS can be used to optimize the allocation of these resources based on real-time demand, supply, and pricing information. The agents in the system can communicate with each other to decide the optimal allocation of resources based on the current conditions of the grid.
2. Real-time monitoring and control: MAS can be used to monitor and control the smart grid in real-time. The agents can analyze data from sensors and other sources to identify potential issues or faults in the grid. Based on this information, the agents can take appropriate actions to avoid or mitigate any problems.

3. Demand response management: MAS can be used to manage demand response programs in a smart grid. The agents can communicate with customers and appliances to manage their energy consumption based on real-time prices and availability. This can help reduce the overall demand on the grid during peak periods, improving its overall efficiency and reliability.
4. Fault detection and isolation: MAS can be used to detect and isolate faults in the smart grid. The agents can analyze data from sensors and other sources to identify potential faults in the grid. Based on this information, the agents can take appropriate actions to isolate the fault and minimize its impact on the rest of the grid.
5. Security and privacy: The smart grid faces challenges in ensuring the security and privacy of the data and communication networks. These challenges can affect the reliability and trustworthiness of the grid. MAS can be used to enhance the security and privacy of the grid by implementing authentication, authorization, and encryption mechanisms. For example, in a MAS-based smart grid, the agents can implement authentication mechanisms to verify the identity of the users and devices.
6. Energy management: The smart grid faces challenges in managing energy consumption, such as peak demand and energy efficiency. These challenges can affect the cost and sustainability of the grid. MAS can be used to manage energy consumption by optimizing the use of energy resources and encouraging consumers to adopt energy-efficient practices. For example, in a MAS-based smart grid, the agents can communicate with consumers to provide them with real-time information about their energy consumption and encourage them to adopt energy-efficient practices. The agents can also optimize the use of energy resources by predicting the demand and supply of energy and adjusting the production and distribution accordingly.
7. Dynamic load management: The smart grid faces challenges in managing dynamic loads, such as electric vehicles and renewable energy sources. These loads can cause fluctuations in the grid's voltage and frequency, which can affect the quality of power supplied to consumers. MAS can be used to manage these dynamic loads by coordinating the charging and discharging of electric vehicles and the integration of renewable energy sources into the grid. For example, in a MAS-based smart grid, electric vehicles can communicate with the grid and with each other to optimize their charging and discharging patterns. The MAS can also optimize the use of renewable energy sources by predicting their availability and integrating them into the grid when they are available.

Overall, the use of MAS can improve the efficiency, reliability, and sustainability of a smart grid by optimizing resource allocation, managing demand response, and detecting and mitigating faults in real-time.

3.1.5.2 Implementation Platforms for MAS

A MAS can be implemented on various platforms, depending on the specific requirements and goals of the system. There is no one-size-fits-all platform for building MAS systems. Instead, developers have a wide range of options to choose from. Some platforms are designed specifically for agent-based modeling and simulation, while others offer more general capabilities for developing complex systems with distributed control and coordination. Regardless of the platform, the central concept in a MAS is the agent. Agents are autonomous entities with their own decision-making capabilities and local knowledge. The platform should provide a suitable level of abstraction to define agents, their behaviors, and how they interact with each other and the environment. Different platforms may support various programming languages. For instance, Java and Python are common choices due to their versatility and extensive libraries for agent development. However, platforms exist for other languages, such as MATLAB or C++, catering to developers with specific language preferences or domain requirements.

Some platforms focus primarily on simulation, enabling developers to model and experiment with agent behaviors before real-world deployment. Other platforms are geared towards real-time control and coordination in operational settings, like industrial automation or energy management systems. Certain platforms are tailored to specific domains, such as smart buildings, energy management, robotics, traffic control, finance, or healthcare. These platforms often come with domain-specific libraries, tools, and models to facilitate the development of agents relevant to that domain. In certain applications, MAS needs to interact with existing systems, devices, or other technologies. Some platforms are designed with built-in support for various communication protocols, making it easier to integrate the MAS with other components of a larger system. Depending on the application, security and privacy may be critical concerns. Some platforms offer features and mechanisms for secure communication and data exchange between agents, protecting sensitive information and preventing unauthorized access. The strength of a platform often lies in its community and support. Active developer communities contribute to the growth and improvement of the platform, providing documentation, examples, and troubleshooting help for other users.

For implementing Multi-Agent Systems (MASs) in the context of smart grids, there are several platforms that offer specific functionalities and capabilities suitable for energy management, optimization, and control in smart grid environments. Here are some popular MAS platforms commonly used for smart grid applications:

1. **JADE (Java Agent Development Framework)**: JADE is a widely used platform for developing MAS applications in Java. It provides a robust infrastructure for building intelligent agents and supports communication protocols commonly used in smart grid systems, such as FIPA-ACL and MQTT. JADE's flexibility and scalability make it suitable for managing distributed energy resources and demand-side management in smart grids.

2. **PowerMatcher**: PowerMatcher is an open-source MAS platform specifically designed for smart grid applications. It is based on JADE and focuses on demand-response and real-time balancing of electricity supply and demand. PowerMatcher allows devices, such as smart appliances and renewable energy sources, to communicate and negotiate in real-time to optimize energy consumption and grid stability.
3. **MAS4J**: MAS4J is an open-source platform based on Java, which provides support for developing energy management and control applications in smart grids. It allows developers to create agents that can interact with various devices, sensors, and other agents to optimize energy use and enhance grid reliability.
4. **NetLogo**: NetLogo is an agent-based modeling platform that is widely used for educational and research purposes. It can also be utilized for developing agent-based simulations of smart grids. NetLogo's user-friendly interface and visualization capabilities make it suitable for understanding the dynamics of energy flows and the impact of various control strategies in smart grids.
5. **AnyLogic**: AnyLogic is a simulation software that supports multi-agent modeling, system dynamics, and discrete-event modeling. It offers a broad range of functionalities, making it applicable to various smart grid scenarios, such as demand response, renewable energy integration, and grid stability analysis.
6. **Repast**: Repast is another general-purpose MAS platform that can be used to simulate complex systems, including smart grids. It provides tools for developing agent-based models of energy markets, grid operations, and consumer behavior to study the impact of different policies and technologies in smart grid environments.

When choosing a MAS platform for smart grid applications, consider factors such as the platform's support for energy-specific communication protocols, its ability to model and simulate the behavior of various grid components, scalability, and the availability of relevant libraries and tools. Additionally, the platform's ease of use and integration with other technologies in the smart grid ecosystem are essential considerations for successful implementation.

3.1.6 Conclusion

The integration of intelligent systems, such as multi-agent systems (MASs), can significantly improve the efficiency and reliability of the smart grid. MAS can be used to address several challenges faced by the smart grid, such as dynamic load management, fault detection and diagnosis, energy management, and security and privacy. They can enable real-time monitoring and control of power generation, transmission, and distribution, and optimize the use of energy resources by predicting the demand and supply of energy. MAS can also encourage consumers to adopt energy-efficient practices and improve the security and privacy of the grid by implementing authentication, authorization, and encryption mechanisms.

As the smart grid continues to evolve and expand, the use of intelligent systems, such as MAS, will become increasingly important in ensuring the efficiency, reliability, and sustainability of the grid. The integration of intelligent systems with the smart grid represents an exciting area of research and development that has the potential to transform the way we generate, transmit, and consume electricity.

The current focus is on increasing the presence of intermittent, low-carbon energy sources and incorporating electric vehicles, heat pumps, and similar loads into smart grids. This poses significant challenges for the grids, including disturbances in power flow and voltage fluctuations that affect both customers and utility equipment. Additionally, ensuring proper operation while minimizing the negative impact of increased energy sources and loads is a difficult task. Coordinated control of smart grids is necessary, and Multi-Agent Systems (MASs) offer a superior solution compared to the existing SCADA systems. Investigating the potential use of MAS for optimized coordinated control of smart grids is essential for improving various complex variables in the networks. MAS is well-suited to manage voltage fluctuations, the smart grid power market, demand-side response, load forecasting, generation forecasting, and generation scheduling during high renewable energy (RE) integration scenarios. Furthermore, MAS can contribute to the realization of the concept of a soft grid. Therefore, implementing MAS for controlling smart grids with a high penetration of RE sources should be considered for future research, aiming to effectively coordinate the aforementioned smart events and processes. Additionally, exploring the application of MAS to manage smart grid flexibility and mitigate RE variability will enhance future smart grid performance and increase the integration of RE sources.

Intelligent systems bring numerous benefits to smart grid management, including enhanced grid optimization, rapid self-recovery, dynamic pricing mechanisms, seamless integration of renewable energy sources, and effective utilization of energy storage devices. These capabilities are crucial for meeting the increasing demands of a growing population while reducing our carbon footprint and promoting sustainability.

Moreover, the coordinated control achieved through intelligent systems enables power utilities to make informed decisions in real-time, ensuring reliable and resilient grid operations. The smart grid's ability to adapt to changing conditions, accommodate renewable energy fluctuations, and respond to consumer demands is critical for building a more resilient and sustainable energy infrastructure.

As technology continues to advance and intelligent systems become more sophisticated, the potential for optimizing smart grid management will only expand further. However, this progress also brings challenges, such as ensuring data security, interoperability, and maintaining the privacy of consumers. Addressing these issues will be crucial to realizing the full potential of intelligent systems in smart grid management.

In conclusion, the integration of intelligent systems marks a significant step forward in our quest for a more efficient, sustainable, and reliable power grid. It is essential for governments, industries, and research institutions to collaborate, innovate, and invest in this transformative technology to build a cleaner, greener, and smarter energy future for generations to come.

References

Albagli AN, Falcão DM, de Rezende JF (2016) Smart grid framework co-simulation using HLA architecture. Electr Power Syst Res 130:22–33

Amirkhani A, Barshooi AH (2022) Consensus in multi-agent systems: a review. Artif Intell Rev 55(5):3897–3935

Bose BK (2017) Artificial intelligence techniques in smart grid and renewable energy systems—some example applications. Proc IEEE 105(11):2262–2273

Colak I, Sagiroglu S, Fulli G, Yesilbudak M, Covrig CF (2016) A survey on the critical issues in smart grid technologies. Renew Sustain Energy Rev 54:396–405

Deng R, Xiao G, Lu R, Chen J (2015) Fast distributed demand response with spatially and temporally coupled constraints in smart grid. IEEE Trans Industr Inf 11(6):1597–1606

Iwayemi A, Yi PZ, Dong XH, Zhou C (2011) Knowing when to act: an optimal stopping method for smart grid demand response. IEEE Netw 25(5):44–49

Järventausta P, Repo S, Rautiainen A, Partanen J (2010) Smart grid power system control in distributed generation environment. Annu Rev Control 34(2):277–286

Judge MA, Khan A, Manzoor A, Khattak HA (2022) Overview of smart grid implementation: Frameworks, impact, performance, and challenges. J Energy Storage 49:104056

Kabalci Y (2016) A survey on smart metering and smart grid communication. Renew Sustain Energy Rev 57:302–318

Le TN, Chin WL, Truong DK, Nguyen TH, Eissa M (2016) Advanced metering infrastructure based on smart meters in smart grid. In: Smart metering technology and services-inspirations for energy utilities

Li F, Qiao W, Sun H, Wan H, Wang J, Xia Y et al (2010) Smart transmission grid: vision and framework. IEEE Trans Smart Grid 1(2):168–177

Li J, Zhao Y, Sun C, Bao X, Zhao Q, Zhou H (2018) A survey of development and application of artificial intelligence in smart grid. In: IOP conference series: earth and environmental science, vol 186. IOP Publishing, Bristol, UK, p 012066

Mahela OP, Khosravy M, Gupta N, Khan B, Alhelou HH, Mahla R et al (2020) Comprehensive overview of multi-agent systems for controlling smart grids. CSEE J Power Energy Syst 8(1):115–131

Manickavasagam K (2014) Intelligent energy control center for distributed generators using multi-agent system. IEEE Trans Power Syst 30(5):2442–2449

Ourahou M, Ayrir W, Hassouni BE, Haddi A (2020) Review on smart grid control and reliability in presence of renewable energies: challenges and prospects. Math Comput Simul 167:19–31

Rubino A, Alessandro S, La Scala M (2021) Handbook of energy economics and policy

Sharma DD, Singh SN, Lin J (2016) Multi-agent based distributed control of distributed energy storages using load data. J Energy Storage 5:134–145

Song EY, FitzPatrick GJ, Lee KB (2017) Smart sensors and standard-based interoperability in smart grids. IEEE Sens J 17(23):7723–7730

Tan KM, Babu TS, Ramachandaramurthy VK, Kasinathan P, Solanki SG, Raveendran SK (2021) Empowering smart grid: a comprehensive review of energy storage technology and application with renewable energy integration. J Energy Storage 39:102591

Wooldridge M (2009) An introduction to multiagent systems. John Wiley & Sons

Xue D, Jing X, Liu H (2019) Detection of false data injection attacks in smart grid utilizing ELM-based OCON framework. IEEE Access 7:31762–31773

Yu M, Hong SH (2015) A real-time demand-response algorithm for smart grids: a Stackelberg game approach. IEEE Trans Smart Grid 7(2):879–888

Chapter 4
Investigation of Electric Load Forecasting Methods: A Weka Application (Regression and Optimization)

Düzgün Akmaz

Abstract Electric load forecasting plays an important role in the planning of power systems. Therefore, it is necessary to develop effective and simple electric load forecasting methods. In this study, a simple and effective method for short-term load forecasting was developed. In the developed method, the highest success with the least input features was aimed. The regression feature of the Random Forest (RF) algorithm and the Correlation Attribute Eval (CAE) feature selection method were used to achieve this. The simulation results proved that the proposed method was successful. While WEKA program was used for RF and CAE algorithms, real-time data was obtained from Spain.

Keywords Electric load forecasting · Feature selection · Random forest · Correlation attribute eval · WEKA

4.1 Introduction to Electric Load Forecasting Methods

In order to use electricity more efficiently, electricity consumption needs to be estimated. Electricity consumption is estimated for certain time intervals using historical data. Different methods were developed in the literature for forecasting electricity consumption. These methods are generally categorized into three groups: time series (Chen et al. 1995; Amral et al. 2007; Saber and Alam 2017; Lynch et al. 2016; Göb et al. 2013; Bercu and Proïa 2013; Charlton and Singleton 2014; Hong et al. 2013), artificial intelligence (Khotanzad et al. 1998; Dong et al. 2021; Yang et al. 2020; Alamaniotis et al. 2016; Johannesen et al. 2018; Khan et al. 2018; Ye et al. 2012; Aribowo et al. 2020; Hong and Wang 2014; Nalcaci et al. 2019), and hybrid methods (Zhang et al. 2018; Wang et al. 2014; Sudheer and Suseelatha 2015; Chaturvedi

D. Akmaz (✉)
Engineering Faculty, Department of Electrical and Electronics Engineering, Munzur University, 62000 Tunceli, Turkey
e-mail: dakmaz@munzur.edu.tr

C. B. Khadse et al. (eds.), *Intelligent Methods in Electrical Power Systems*, Engineering Optimization: Methods and Applications, https://doi.org/10.1007/978-981-97-5718-3_4

et al. 2015; Yu et al. 2021; Li et al. 2016; Tayab et al. 2020; El-Hendawi and Wang 2020; Abdoos et al. 2015; Fard and Akbari-Zadeh 2014; Capuno et al. 2017).

A method based on an Adaptive Time-Series Autoregressive Moving Average (ARMA) was developed for short-term load forecasting (Chen et al. 1995). This study compared the ARMA method with the Box-Jenkins approach. The results obtained showed that the ARMA method was more successful (Chen et al. 1995). A method based on Multiple Linear Regression (MLR) was developed (Amral et al. 2007). In the method, historical temperature data and load demand were evaluated. On the other hand, the MLR method was applied to big data (Saber and Alam 2017). Multi-core parallel processing was applied for MLR (Saber and Alam 2017). An expanded Kalman filter bank methodology was developed for short-term load forecasting (Lynch et al. 2016). Exponential smoothing with covariates was used (Göb et al. 2013). Multi-seasonality was considered in the exponential smoothing model (Göb et al. 2013). A Sarimax model (Bercu and Proïa 2013) and a refined parametric model (Charlton and Singleton 2014) were also used for short-term load forecasting. Finally, a long-term load forecasting method was developed using the MLR model and the weather normalization method (Hong et al. 2013).

As can be seen from (Chen et al. 1995; Amral et al. 2007; Saber and Alam 2017; Lynch et al. 2016; Göb et al. 2013; Bercu and Proïa 2013; Charlton and Singleton 2014; Hong et al. 2013), different time series methods were developed for load forecasting. However, since the electric load is affected by many different parameters, time series methods may not give good results in some cases. Therefore, different Artificial Intelligence (AI) methods were used in different studies for electric load forecasting (Khotanzad et al. 1998; Dong et al. 2021; Yang et al. 2020; Alamaniotis et al. 2016; Johannesen et al. 2018; Khan et al. 2018; Ye et al. 2012; Aribowo et al. 2020; Hong and Wang 2014; Nalcaci et al. 2019). A model consisting of using multiple Artificial Neural Networks (ANN) was proposed (Khotanzad et al. 1998). A method based on the Bagged Regression Trees (BRT) model was applied (Dong et al. 2021). The speed and accuracy of the method were compared with ANN. The results showed that the BRT method was more accurate and faster than ANN (Dong et al. 2021). A novel Deep Learning (DL) approach was developed for short- and medium-term electrical load (Yang et al. 2020). It was stated that DL methods can be effective for uncertain and large amounts of data (Yang et al. 2020). Gaussian process regression and relevance vector regression were used to forecast future load demand (Alamaniotis et al. 2016). Random Forest (RF) and K-Nearest Neighbors algorithm (kNN) were used for short- and long-term electric load forecasting (Johannesen et al. 2018). The results of the study showed that RF was more successful with 30 min of data, while kNN was more successful with 24 h of data (Johannesen et al. 2018). Similar to the study (Dong et al. 2021), ANN and BRT were compared (Khan et al. 2018). BRT was reported to be more effective than ANN (Khan et al. 2018). Support Vector Regression (SVR) was used for long-term load forecasting (Ye et al. 2012). Generalized regression neural network was used to predict long-term electricity load (Aribowo et al. 2020). Fuzzy interaction regression was used (Hong and Wang 2014). Multivariate Adaptive Regression Splines (MARS), ANN, and linear regression methods were compared (Nalcaci et al. 2019). The results of the study showed that the MARS model was more effective for long-term load forecasting (Nalcaci et al. 2019). In general, artificial intelligence methods need large amounts of data for training and testing.

Finally, hybrid methods were used for electric load forecasting (Zhang et al. 2018; Wang et al. 2014; Sudheer and Suseelatha 2015; Chaturvedi et al. 2015; Yu et al. 2021; Li et al. 2016; Tayab et al. 2020; El-Hendawi and Wang 2020; Abdoos et al. 2015; Fard and Akbari-Zadeh 2014; Capuno et al. 2017; Akmaz 2022). Hybrid methods include the use of more than one approach. Therefore, hybrid methods are mathematically complex. Improved empirical mode decomposition, autoregressive integrated moving average, and Wavelet Neural Networks (WNN) were used (Zhang et al. 2018). In this study, WNN was optimized with the fruit fly optimization algorithm (Zhang et al. 2018). A hybrid method based on the Seasonal Autoregressive Integrated Moving Average Mode (SARIMA) and Back Propagation Neural Network (BPNN) was developed (Wang et al. 2014). Daubechies wavelet of order 3 (db3) was used to reduce the noise effect (Wang et al. 2014). The low-frequency component obtained after Wavelet Transform (WT) was used for SARIMA and BPNN (Wang et al. 2014). Estimates of SARIMA and BPNN were combined with the variance–covariance technique (Wang et al. 2014). WT, triple exponential smoothing model, and weighted nearest neighbor were used (Sudheer and Suseelatha 2015). A Haar wavelet filter was used to decompose the load signal into its deterministic and fluctuation components (Sudheer and Suseelatha 2015). WT, an adaptive genetic algorithm, and a fuzzy system with a generalized neural network were used (Chaturvedi et al. 2015). A generative adversarial network, autoregressive integrated moving average, and wavelet package were used to estimate short-term load forecasting (Yu et al. 2021). The WT-extreme learning machine and partial least squares regression were used (Li et al. 2016). Different mother wavelets such as Daubechies (db2–db5) and Coiflets (coif2–coif5) were used (Li et al. 2016). Signals were divided into three levels (Li et al. 2016). The best-basis stationary Wavelet Packet Transform (WPT) and Harris Hawks Optimization (HHO)-based feed-forward neural network was used (Tayab et al. 2020). Full WPT and neural networks were used (El-Hendawi and Wang 2020). WT-Gram–Schmidt feature selection method and support vector machine were used (Abdoos et al. 2015). WT, Autoregressive Integrated Moving Average (ARIMA), and ANN were applied (Fard and Akbari-Zadeh 2014). A hybrid method based on algebraic prediction and SVR was developed for very short-term load forecasting (Capuno et al. 2017). Multilayer perceptron algorithm, correlation-based feature selection method, and curve fitting techniques were used for long-term load forecasting (Akmaz 2022). In Table 4.1, the studies reviewed in this study were compared in some respects.

The methods in the literature have some features (Khan et al. 2016; Hong and Fan 2016; Raza and Khosravi 2015; Kuster et al. 2017). These features are summarized in (Khan et al. 2016; Hong and Fan 2016; Raza and Khosravi 2015; Kuster et al. 2017).

In this study, after reviewing some of the methods in the literature, a short-term load forecasting method was also developed using real-time data from Spain and the WEKA program. In the developed method, the regression feature of the Random

Table 4.1 Comparison of different studies in some respects

Refs.	Model	Signal processing	Optimization	Type	Data
Chen et al. (1995)	Time	–	–	Short term	The load of Taipower system
Amral et al. (2007)	Time	–	–	Short term	The South Sulawesi Network
Saber and Alam (2017)	Time	–	–	Short term	–
Lynch et al. (2016)	Time	–	–	Short term	Cork Institute of Technology (CIT)
Göb et al. (2013)	Time	–	–	Short term	Emilia-Romagna/ Italian region
Hong et al. (2013)	Time	–	–	Long term	North Carolina Electric Membership Corporation
Khotanzad et al. (1998)	Artificial intelligence	–	–	Short term	Texas Utilities Electric (TUE) Control Center in Dallas
Dong et al. (2021)	Artificial intelligence	–	–	Short term	Qingdao/China
Yang et al. (2020)	Artificial intelligence	–	–	Short/ medium-term electrical	–
Alamaniotis et al. (2016)	Artificial intelligence	–	–	Medium term electricity	ISO/New England http://www.iso-ne.com/
Johannesen et al. (2018)	Artificial intelligence	–	–	Short term	Australian Energy Market Operator (AEMO) Bureau of Meteorology (BOM) Sydney/ NSW region
Khan et al. (2018)	Artificial intelligence	–	–	Short term	Islamabad Electric Supply Company (IESCO), Pakistan
Ye et al. (2012)	Artificial intelligence	–	–	Long term	China
Aribowo et al. (2020)	Artificial intelligence	–	–	Long-term electricity	Span
Hong and Wang (2014)	Artificial intelligence	–	–	Short term	ISO/New England

(continued)

Table 4.1 (continued)

Refs.	Model	Signal processing	Optimization	Type	Data
Nalcaci et al. (2019)	Artificial intelligence	–	–	Long term	Turkish Electricity Transmission Company Turkish State Meteorological Service
Zhang et al. (2018)	A hybrid model	Improved empirical mode decomposition	Fruit fly	Short-term electricity	*Australian The electricity load data https://www.aemo.com.au Temperature data https://www.bom.gov.au/ *New York City The electricity load data www.nyiso.com Temperature data www.weather.gov
Wang et al. (2014)	A hybrid model	Wavelet transform	–	Short-term electricity	New South Wales/Australia
Sudheer and Suseelatha (2015)	A hybrid model	Wavelet transform	–	Short-term electricity	The electricity markets of California http://www.ucei.berkely.edu/ucei/datamine/ucei-data Spain http://www.omel.es/
Chaturvedi et al. (2015)	A hybrid model	Wavelet transform	–	Short-term electricity	Dayalbagh Educational Institute, Dayalbagh, Agra/India
Yu et al. (2021)	A hybrid model	Wavelet package	–	Short term	Shiraz Electricity Company/Iran
Li et al. (2016)	A hybrid model	Wavelet transform	–	Short term	ISO/New England http://www.iso-ne.com/isoexpress/web/reports/pricing/-/tree/zone-info

(continued)

Table 4.1 (continued)

Refs.	Model	Signal processing	Optimization	Type	Data
Tayab et al. (2020)	A hybrid model	Stationary wavelet packet transform	Harris hawks optimization	Short term	The load demand: Queensland grid Weather data: Griffith University (Nathan campus)
El-Hendawi and Wang (2020)	A hybrid model	Full wavelet packet transform	–	Short term	Independent electricity system operator (IESO)/ Canada http://www.ieso.ca/%20power-data The temperature and humidity https://climate.weather.gc.ca
Abdoos et al. (2015)	A hybrid model	Wavelet transform	–	Short term	Iran's load and weather information http://sccisrep.igmc.ir/RptParam.aspx http://www.wunderground.com/history/m
Fard and Akbari-Zadeh (2014)	A hybrid model	Wavelet transform	–	Short term	Fars Electric Power Company/ Iran http://www.frec.co.ir/
Capuno et al. (2017)	A hybrid model	–	–	Very short term	South Korea
Akmaz (2022)	A hybrid model	–	Correlation-based feature selection method	Long-term total electricity	Turkey TUİK https://www.tuik.gov.tr/
Pro	Artificial intelligence	–	Correlation attribute eval	Short term	Spain https://www.neuraldesigner.com/blog/electricity_demand_forecasting

Forest (RF) algorithm and the Correlation Attribute Eval (CAE) feature selection method were used. The simulation results proved that the proposed method was successful.

Detailed information about the RF used in this study can be found in (Breiman 2001; Khoshgoftaar et al. 2007; Guo et al. 2004). Also, detailed information about CAE can be found in (https://weka.sourceforge.io/doc.dev/weka/attributeSelection/CorrelationAttributeEval.html). The developed method was described in detail in Sects. 4.3 and 4.4.

4.2 Real-Time Data Information

In this study, real-time data were obtained from Spain. This database can be found (https://www.neuraldesigner.com/blog/electricity_demand_forecasting). This database contains electricity consumption data for Spain between 2015 and 2018 years. At the same time, this database contains different weather data for five cities in Spain (Barcelona, Bilbao, Madrid, Seville, and Valencia). Table 4.2 summarizes the names of the variables included in this dataset. Only January data was used in this study.

The adjusted January data consists of 14,845 different rows. A randomly selected 80% of the January data were used for training. The remaining 20% data was used for testing.

4.3 WEKA Application (Regression)

In this study, the WEKA program was used to implement the regression features of different artificial intelligence methods. Figure 4.1 shows the WEKA program.

First, the data needs to be entered into the WEKA program. The input format of the data in the WEKA program is shown in Fig. 4.2.

After the data was entered into WEKA, the regression feature of different algorithms such as RF, Instance-Based Learner algorithm (IBK), Decision Table (DT), and SMOreg methods can be used in the WEKA program. Figure 4.3 shows some of the regression results obtained with RF in the WEKA program.

The Mean Absolute Percentage Error (MAPE) is calculated by Eq. (4.1).

$$\text{MAPE} = \frac{\sum_{i=1}^{n} \left| \frac{(X_i - Z_i)}{Z_i} \right|}{n} * 100 \tag{4.1}$$

Here, X_i represents the predicted value and Z_i represents the actual value. n represents the total number of data.

MAPE results of RF, IBK, DT, and SMOreg algorithms are given in Table 4.3.

Table 4.2 Spain database names and features (January)

Inputs	Output
Name	
1. Year	Electricity_consumption
2. Day	
3. Hour	
4. City_name	
5. Clouds_all	
6. Humidity	
7. Pressure	
8. Rain_1h	
9. Rain_3h	
10. Snow_3h	
11. Temperature	
12. Temperaturemax	
13. Temperaturemin	
14. Weather_description	
15. Weather_icon	
16. Weather_id	
17. Weather_main	
18. Wind_deg	
19. Wind_speed	

Actual values and estimated values should be as close to each other as possible. This is determined by the minimum MAPE value. A minimum MAPE value indicates that the method was successful.

As seen in Table 4.3, the lowest MAPE value in the test data was obtained with RF. Therefore, the use of the RF algorithm was proposed in this study. In Table 4.4, the times used by the RF algorithm for training and testing were given.

4.4 WEKA Application (Regression and Optimization)

The results in Tables 4.3 and 4.4 were obtained with 19 input features. Some of these features may not contain effective information for training/testing regression results. More successful training and testing result can be obtained with some determined features. Effective features can be determined using the WEKA program. In this study, it was determined that CAE was effective for feature selection. Therefore, the CAE method was used to rank the effective properties. The effective feature values for regression according to the CAE method are given in Fig. 4.4.

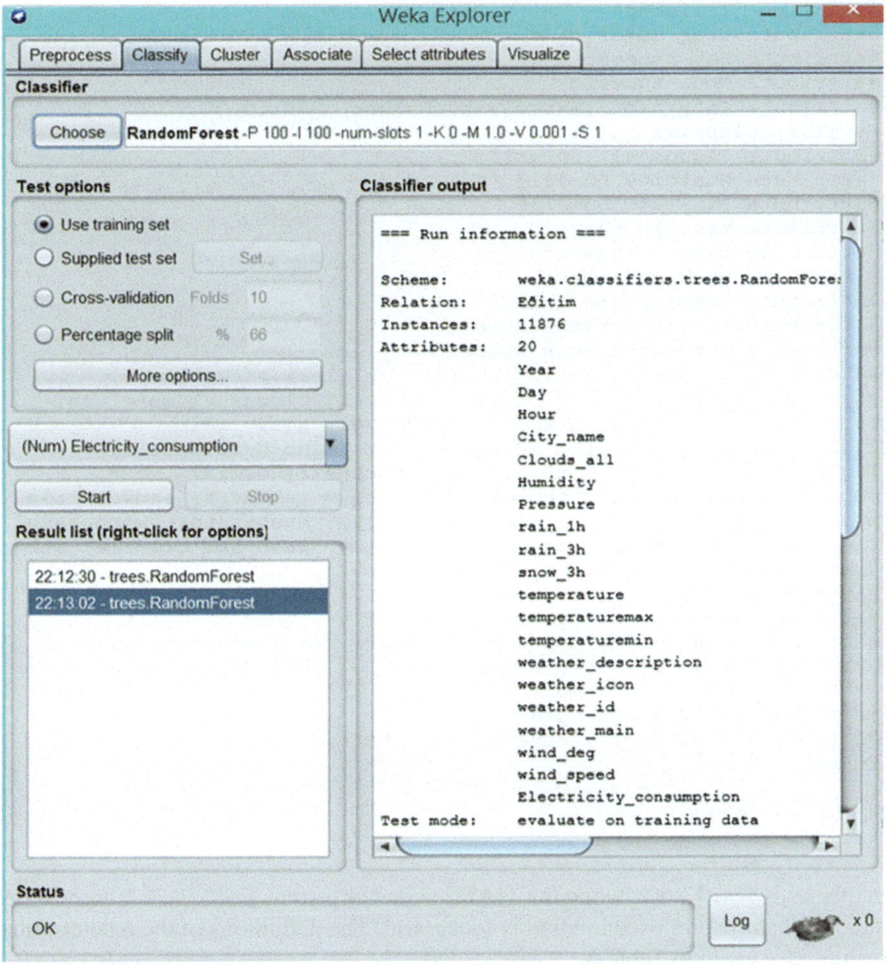

Fig. 4.1 WEKA program

According to the CAE method, the most effective input variable for short-term electricity consumption was the hour. The least effective variable was the humidity. These rankings may vary depending on the database and the feature selection method. In CAE, a ranking value closer to 1 indicates that the input variable is directly related to electricity consumption. After the features were listed, training and testing success with the RF algorithm was examined. Table 4.5 shows the training and testing MAPE values obtained by adding sequential features.

As seen in Table 4.5, the highest training and testing success was obtained with 6 features determined by RF and CAE. In the literature, it is stated that a successful estimation is achieved if the MAPE is less than 10% (Aribowo et al. 2020; ES et al. 2014; Luo et al. 2018). As can be seen from Table 4.5, the MAPE value for training

```
@relation 'Training'
@attribute Year numeric
@attribute Day numeric
@attribute Hour numeric
@attribute City_name {Barcelona, Bilbao, Madrid,
Seville,Valencia}
@attribute Clouds_all numeric
@attribute Humidity numeric
@attribute Pressure numeric
@attribute rain_1h numeric
@attribute rain_3h numeric
@attribute snow_3h numeric
@attribute temperature numeric
@attribute temperaturemax numeric
@attribute temperaturemin numeric
@attribute weather_description {sky_is_clear, few_clouds,
scattered_clouds, broken_clouds, overcast_clouds, light_rain,
moderate_rain, heavy_intensity_rain, very_heavy_rain,
light_intensity_drizzle, drizzle, mist, sand_dust_whirls,
light_intensity_shower_rain, light_intensity_drizzle_rain,
shower_rain, proximity_shower_rain, heavy_intensity_shower_rain,
proximity_thunderstorm, fog, light_snow, thunderstorm_with_rain,
light_rain_and_snow, rain_and_snow, haze, rain_and_drizzle, snow,
dust, thunderstorm_with_light_rain}
@attribute weather_icon numeric
@attribute weather_id numeric
@attribute weather_main {clear, clouds, rain, drizzle, mist,
dust, thunderstorm, fog, snow, haze}
@attribute wind_deg numeric
@attribute wind_speed numeric
@attribute Electricity_consumption numeric
@data
```

Fig. 4.2 WEKA input format

in this study was 1.86%, while the MAPE value for testing was 4.94%. These values indicate that the proposed method is successful. The definitions of the 6 determined features are given in Table 4.6.

Since 6 features were determined with the CAE and RF algorithm, the training and testing time was expected to be shorter. The time required for training and testing for the 6 features is given in Table 4.7.

When Table 4.4 (19-features) and Table 4.7 (6-features) were examined, it was seen that the training and testing periods were shorter with 6 features. With this approach, a more efficient short-term electricity consumption forecasting method was developed with the least input feature. The time required for training and testing was reduced due to less use of input features.

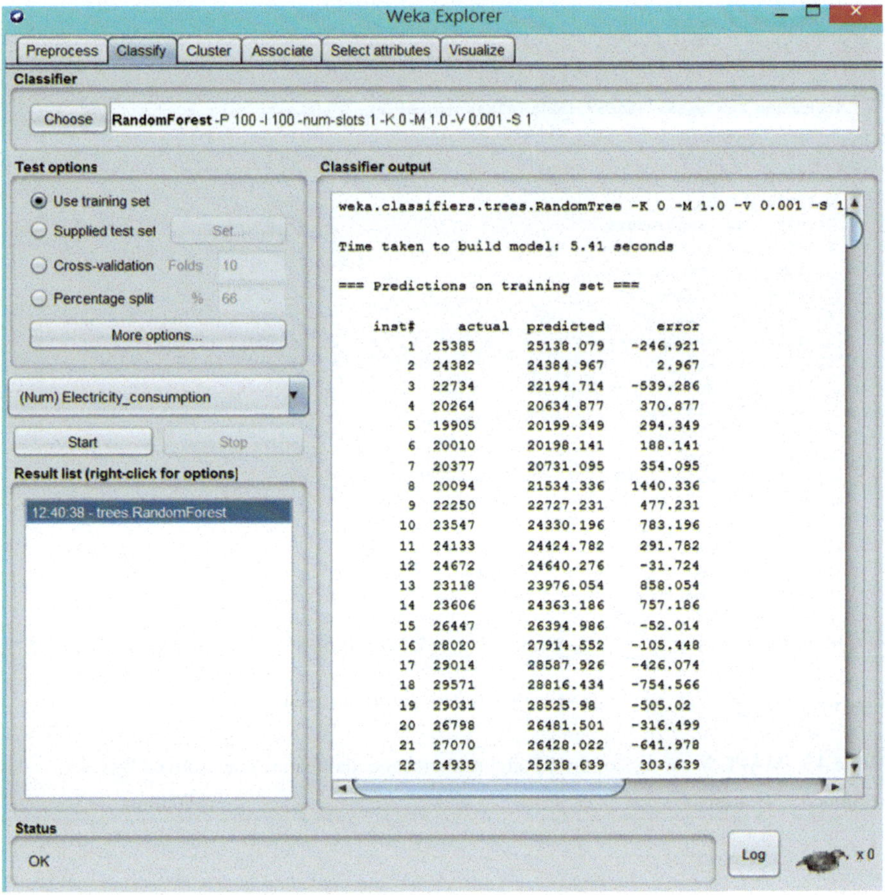

Fig. 4.3 Random forest regression results

Table 4.3 MAPE % values of different algorithms (19 features-Table 4.2)

Algorithm	Training	Testing
Random forest	1.91	5.26
Decision table	6.52	7.66
Instance-based learner algorithm (IBK)	0	8.10
SMOreg	11.52	11.90

Table 4.4 Training and testing times of the RF algorithm (19 features-Table 4.2)

Random forest	Training (s)	Testing (s)
Time taken to build model	7.15	6.47
Time taken to test model on training data	3.91	1.03

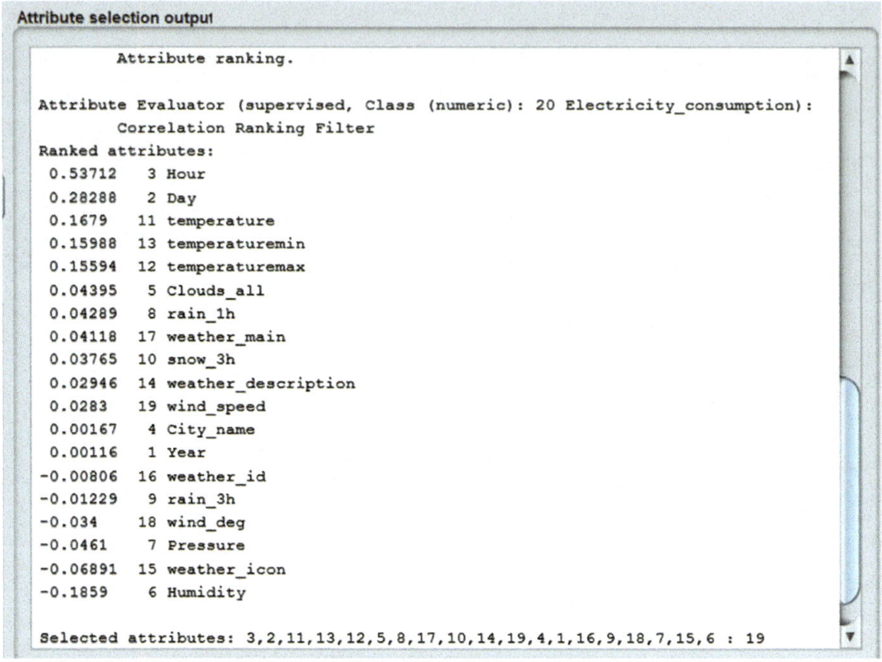

```
Attribute selection output

        Attribute ranking.

Attribute Evaluator (supervised, Class (numeric): 20 Electricity_consumption):
        Correlation Ranking Filter
Ranked attributes:
 0.53712    3 Hour
 0.28288    2 Day
 0.1679    11 temperature
 0.15988   13 temperaturemin
 0.15594   12 temperaturemax
 0.04395    5 Clouds_all
 0.04289    8 rain_1h
 0.04118   17 weather_main
 0.03765   10 snow_3h
 0.02946   14 weather_description
 0.0283    19 wind_speed
 0.00167    4 City_name
 0.00116    1 Year
-0.00806   16 weather_id
-0.01229    9 rain_3h
-0.034     18 wind_deg
-0.0461     7 Pressure
-0.06891   15 weather_icon
-0.1859     6 Humidity

Selected attributes: 3,2,11,13,12,5,8,17,10,14,19,4,1,16,9,18,7,15,6 : 19
```

Fig. 4.4 Attribute ranking obtained with correlation attribute eval

Table 4.5 MAPE % success of the RF algorithm (sequential features determined by CAE)

Number	Features	Training	Testing
1	3	10.03	10.36
2	3, 2	5.85	6.14
3	3, 2, 11	2.19	5.89
4	3, 2, 11, 13	1.98	5.35
5	3, 2, 11, 13, 12	1.93	5.17
6	**3, 2, 11, 13, 12, 5**	**1.86**	**4.94**
7	3, 2, 11, 13, 12, 5, 8	1.86	5.08
8	3, 2, 11, 13, 12, 5, 8, 17	1.88	4.94
9	3, 2, 11, 13, 12, 5, 8, 17, 10	1.91	5.02
10	3, 2, 11, 13, 12, 5, 8, 17, 10, 14	2.06	5.48
11	3, 2, 11, 13, 12, 5, 8, 17, 10, 14, 19	2.07	5.54
12	3, 2, 11, 13, 12, 5, 8, 17, 10, 14, 19, 4	2.09	5.70
...	...	...	...
19	3, 2, 11, 13, 12, 5, 8, 17, 10, 14, 19, 4, 1, 16, 9, 18, 7, 15, 6	1.91	5.26

Table 4.6 Definitions of features determined by CAE and RF algorithm

Number	Input	Input variable name
1	3	Hour
2	2	Day
3	11	Temperature
4	13	Temperaturemin
4	12	Temperaturemax
6	5	Clouds_all

Table 4.7 Training and testing times of the RF algorithm (6 features-Table 4.6)

RF	Training (s)	Testing (s)
Time taken to build model	5.28	5.5
Time taken to test model on training data	3.88	1

4.5 Summary

Electric load forecasting is one of the most important issues in power systems. Therefore many research/methods were completed to estimate electricity consumption. In this study, the methods developed for the forecasting of electricity consumption were investigated. At the same time, a short-term electric forecasting method was developed in this study.

The regression feature of the RF algorithm was used to estimate the electrical load consumption. It has been determined that RF is successful in estimating electrical load forecasting values for the database. However, the 19 different input features in the database cause a long training and testing period. Therefore, it is necessary to select effective input features and make predictions of the RF algorithm with less input features. The regression feature of the RF algorithm and the CAE feature selection method were used to perform this. Firstly, the CAE method was used to rank the effective features. Then, Electric Load Forecasting values were determined by using the selected input features and the RF algorithm. The simulation results showed that a higher successful prediction could be made with the selected less input feature. As the number of input features decreased, the time required for RF training and testing decreased. The simulation results showed that the applied method was successful.

For future work

- Different databases can be used.
- Different artificial intelligence methods can be used.
- Different feature selection methods can be used.
- Hybrid methods can be developed.

References

Abdoos A, Hemmati M, Abdoos AA (2015) Short term load forecasting using a hybrid intelligent method. Knowl-Based Syst 76:139–147. https://doi.org/10.1016/j.knosys.2014.12.008

Akmaz D (2022) Çok katmanlı algılayıcı algoritması, korelasyon tabanlı özellik seçme yöntemi ve eğri uydurma tekniği ile türkiye'de toplam elektrik tüketiminin tahmin edilmesi. Fırat Üniversitesi Mühendislik Bilimleri Dergisi 34(2):677–686. https://doi.org/10.35234/fumbd.1118732

Alamaniotis M, Bargiotas D, Tsoukalas LH (2016) Towards smart energy systems: application of kernel machine regression for medium term electricity load forecasting. Springerplus 5(1):1–15. https://doi.org/10.1186/s40064-016-1665-z

Amral N, Ozveren CS, King D (2007) Short term load forecasting using multiple linear regression. In: 42nd IEEE international universities power engineering conference, pp 1192–1198. https://doi.org/10.1109/UPEC.2007.4469121

Aribowo W, Muslim S, Basuki I (2020) Generalized regression neural network for long-term electricity load forecasting. In: International conference on smart technology and applications (ICoSTA), pp 1–5. https://doi.org/10.1109/ICoSTA48221.2020.1570611361

Bercu S, Proïa F (2013) A SARIMAX coupled modelling applied to individual load curves intraday forecasting. J Appl Stat 40(6):1333–1348. https://doi.org/10.1080/02664763.2013.785496

Breiman L (2001) Random forests. Mach Learn 45(1):5–32. https://doi.org/10.1023/A:1010933404324

Capuno M, Kim JS, Song H (2017) Very short-term load forecasting using hybrid algebraic prediction and support vector regression. Math Probl Eng 8298531. https://doi.org/10.1155/2017/8298531

Charlton N, Singleton C (2014) A refined parametric model for short term load forecasting. Int J Forecast 30(2):364–368. https://doi.org/10.1016/j.ijforecast.2013.07.003

Chaturvedi DK, Sinha AP, Malik OP (2015) Short term load forecast using fuzzy logic and wavelet transform integrated generalized neural network. Int J Electr Power Energy Syst 67:230–237. https://doi.org/10.1016/j.ijepes.2014.11.027

Chen JF, Wang WM, Huang CM (1995) Analysis of an adaptive time-series autoregressive moving-average (ARMA) model for short-term load forecasting. Electr Power Syst Res 34(3):187–196. https://doi.org/10.1016/0378-7796(95)00977-1

Dong H, Gao Y, Fang Y, Liu M, Kong Y (2021) The short-term load forecasting for special days based on bagged regression trees in Qingdao, China. Comput Intell Neurosci 3693294. https://doi.org/10.1155/2021/3693294

El-Hendawi M, Wang Z (2020) An ensemble method of full wavelet packet transform and neural network for short term electrical load forecasting. Electr Power Syst Res 182:106265. https://doi.org/10.1016/j.epsr.2020.106265

Ha ES, Kalender FY, Hamzaçebi C (2014) Yapay sinir ağlari ile Türkiye net enerji talep tahmini. Gazi Üniversitesi Mühendislik-Mimarlık Fakültesi Dergisi 29(3):495–504

Fard AK, Akbari-Zadeh MR (2014) A hybrid method based on wavelet, ANN and ARIMA model for short-term load forecasting. J Exp Theor Artif Intell 26(2):167–182. https://doi.org/10.1080/0952813X.2013.813976

Göb R, Lurz K, Pievatolo A (2013) Electrical load forecasting by exponential smoothing with covariates. Appl Stoch Model Bus Ind 29(6):629–645. https://doi.org/10.1002/asmb.2008

Guo L, Ma Y, Cukic B, Singh H (2004) Robust prediction of fault-proneness by random forests. In: 15th international symposium on software reliability engineering, pp 417–428. https://doi.org/10.1109/ISSRE.2004.35

Hong T, Fan S (2016) Probabilistic electric load forecasting: a tutorial review. Int J Forecast 32(3):914–938. https://doi.org/10.1016/j.ijforecast.2015.11.011

Hong T, Wang P (2014) Fuzzy interaction regression for short term load forecasting. Fuzzy Optim Decis Making 13(1):91–103. https://doi.org/10.1007/s10700-013-9166-9

Hong T, Wilson J, Xie J (2013) Long term probabilistic load forecasting and normalization with hourly information. IEEE Trans Smart Grid 5(1):456–462. https://doi.org/10.1109/TSG.2013.2274373

https://weka.sourceforge.io/doc.dev/weka/attributeSelection/CorrelationAttributeEval.html. Accessed November 2022

https://www.neuraldesigner.com/blog/electricity_demand_forecasting. Accessed November 2022

Johannesen NJ, Kolhe M, Goodwin M (2018) Comparison of regression tools for regional electric load forecasting. In: 3rd international conference on smart and sustainable technologies (SpliTech), pp 1–6

Khan AR, Mahmood A, Safdar A, Khan ZA, Khan NA (2016) Load forecasting, dynamic pricing and DSM in smart grid: A review. Renew Sustain Energy Rev 54:1311–1322. https://doi.org/10.1016/j.rser.2015.10.117

Khan AR, Razzaq S, Alquthami T, Moghal MR, Amin A, Mahmood A (2018) Day ahead load forecasting for IESCO using artificial neural network and bagged regression tree. In: 1st international conference on power, energy and smart grid (ICPESG), pp 1–6. https://doi.org/10.1109/ICPESG.2018.8384489

Khoshgoftaar TM, Golawala M, Van Hulse J (2007) An empirical study of learning from imbalanced data using random forest. In: 19th IEEE international conference on tools with artificial intelligence (ICTAI 2007), pp 310–317. https://doi.org/10.1109/ICTAI.2007.46

Khotanzad A, Afkhami-Rohani R, Maratukulam D (1998) ANNSTLF-artificial neural network short-term load forecaster-generation three. IEEE Trans Power Syst 13(4):1413–1422.https://doi.org/10.1109/59.736285

Kuster C, Rezgui Y, Mourshed M (2017) Electrical load forecasting models: a critical systematic review. Sustain Cities Soc 35:257–270. https://doi.org/10.1016/j.scs.2017.08.009

Li S, Goel L, Wang P (2016) An ensemble approach for short-term load forecasting by extreme learning machine. Appl Energy 170:22–29. https://doi.org/10.1016/j.apenergy.2016.02.114

Luo J, Hong T, Fang SC (2018) Robust regression models for load forecasting. IEEE Trans Smart Grid 10(5):5397–5404. https://doi.org/10.1109/TSG.2018.2881562

Lynch C, JO'Mahony M, Guinee RA (2016) Electrical load forecasting using an expanded Kalman filter bank methodology. IFAC-PapersOnLine 49(25):358–365. https://doi.org/10.1016/j.ifacol.2016.12.069

Nalcaci G, Özmen A, Weber GW (2019) Long-term load forecasting: models based on MARS, ANN and LR methods. CEJOR 27(4):1033–1049. https://doi.org/10.1007/s10100-018-0531-1

Raza MQ, Khosravi A (2015) A review on artificial intelligence based load demand forecasting techniques for smart grid and buildings. Renew Sustain Energy Rev 50:1352–1372. https://doi.org/10.1016/j.rser.2015.04.065

Saber AY, Alam AR (2017) Short term load forecasting using multiple linear regression for big data. In: IEEE symposium series on computational intelligence, pp 1–6. https://doi.org/10.1109/SSCI.2017.8285261

Sudheer G, Suseelatha A (2015) Short term load forecasting using wavelet transform combined with Holt-Winters and weighted nearest neighbor models. Int J Electr Power Energy Syst 64:340–346. https://doi.org/10.1016/j.ijepes.2014.07.043

Tayab UB, Zia A, Yang F, Lu J, Kashif M (2020) Short-term load forecasting for microgrid energy management system using hybrid HHO-FNN model with best-basis stationary wavelet packet transform. Energy 203:117857. https://doi.org/10.1016/j.energy.2020.117857

Wang J, Wang J, Li Y, Zhu S, Zhao J (2014) Techniques of applying wavelet de-noising into a combined model for short-term load forecasting. Int J Electr Power Energy Syst 62:816–824. https://doi.org/10.1016/j.ijepes.2014.05.038

Yang Y, Haq EU, Jia Y (2020) A novel deep learning approach for short and medium-term electrical load forecasting based on pooling LSTM-CNN model. In: IEEE/IAS industrial and commercial power system Asia (I&CPS Asia), pp 26–34. https://doi.org/10.1109/ICPSAsia48933.2020.9208557

Ye S, Zhu G, Xiao Z (2012) Long term load forecasting and recommendations for China based on support vector regression. Energy Power Eng 4(05):380–385. https://doi.org/10.4236/epe.2012.45050

Yu F, Yue Q, Yunianta A, Aljahdali HMA (2021) A novel hybrid deep correction approach for electrical load demand prediction. Sustain Cities Soc 74:103161. https://doi.org/10.1016/j.scs.2021.103161

Zhang J, Wei YM, Li D, Tan Z, Zhou J (2018) Short term electricity load forecasting using a hybrid model. Energy 158:774–781. https://doi.org/10.1016/j.energy.2018.06.012

Chapter 5
Scaled Conjugate Gradient-Based Intelligent Microgrid Fault Analysis System

Nishant Chaudhari, Sharvari Dhamale, Anand Mahajan, and Chetan B. Khadse

Abstract A microgrid fed small three-phase transmission line under various fault conditions is analyzed in this paper with the help of artificial neural network. The transmission line model along with different fault conditions are simulated in MATLAB Simulink. The simulation setup includes three sources, viz., PV solar source, battery, and 3-phase AC source. These sources are followed by inverters, transformer, RC filters, circuit breakers, and a three-phase transmission line model. The three-phase inverter converts DC solar and battery inputs to three-phase AC. The transformer elevates the three-phase AC voltage. RC filters eliminate harmonics generated due to power electronics equipment. The circuit breaker switches between sources so that various fault conditions can be generated with respect to magnitude and time. This fault data is used as an input to train the neural network. The neural network is trained with symmetrical and asymmetrical fault conditions for various duration and time. Some of the samples from generated fault conditions are used to test the neural network. The results are used to comment on accuracy and feasibility of the system.

Keywords Artificial intelligence · Micro grid · Neural network · Pattern recognition · Renewable energy

5.1 Introduction

The microgrid is gaining much attention in some recent years. It may prove the perfect balance between non-conventional energy sources and conventional energy sources. Both the sources can be utilized with the perfect optimization. As it is much beneficial

N. Chaudhari · S. Dhamale · C. B. Khadse (✉)
Dr. Vishwanath Karad MIT World Peace University, Pune, Maharashtra, India
e-mail: chetan.khadse@mitwpu.edu.in

A. Mahajan
Delft University of Technology, Delft, The Netherlands

C. B. Khadse et al. (eds.), *Intelligent Methods in Electrical Power Systems*, Engineering Optimization: Methods and Applications, https://doi.org/10.1007/978-981-97-5718-3_5

for the humankind, analysis of such systems before using is always necessary. As regular transmission lines are subjected to the faults, the microgrid transmission line may also have the same danger. It is important to study the effect of symmetrical and unsymmetrical faults on the microgrid transmission lines. There is a possibility that the microgrid will collapse due to various symmetrical and unsymmetrical faults conditions. To keep it in operating condition, these defects need to be fixed at a specific time. The bulk of overhead transmission lines are extremely vulnerable to extreme weather and harsh environments, resulting in failures. The precise location of a transmission line problem spot can reduce the number of persons and materials needed for line inspection following a fault resulting in an improvement in power supply reliability.

The traditional fault analysis approaches use wavelet transform, Stockwell transform, fuzzy logic, artificial neural network, support vector machine, etc. Out of which use of artificial neural network for the fault detection and classification is proven accurate by many researchers. There are various training algorithms that have been studied over past decades for the classification of faults (Khadke et al. 1834). The artificial neural network is used for the low latency of fault detection and classification in Ogar et al. (2023). The 11 kV transmission line's fault protection system is proposed in Pandey (2021). Likewise, many literature papers have proven the benefits of neural network.

The microgrid protection is a critical problem which is addressed and elaborated in Singh et al. (2020). The artificial neural network is proven accurate in case of microgrid also. Many learning algorithms have been used to train the neural network in case of microgrid fault analysis. An intelligent fault detection system for microgrid based on ANN is proposed in Esmaeilbeigi and Karegar (2020). The high impedance fault detection based on ANN in microgrids is explained in Grcić and Pandžić (2023). Electromagnetic field and neural network-based fault detection system for smartgrid is proposed in Khadse et al. (2021).

In this paper, scaled conjugate gradient descent back-propagation (SCGB) algorithm is used as a learning algorithm. This algorithm is proven to be the best among other back-propagation algorithm when considered for power quality analysis (Khadse et al. June 2017; Khadse et al. 2017). This algorithm is also used to design the pattern classifier for chemical compounds in agarwood (Zubir et al. 2017). The RADAR pulse modulation estimator is developed with the SCGB algorithm in Lunden and Koivunen (2007). The automated seizer detector is created with the help of SCGB algorithm in Sivasankari et al. (2013).

An attempt is made here to check the accuracy of algorithm for microgrid fault detection and classification. The paper is organized as follows: Second section explains the proposed methodology. Third section will elaborate the procedure of the experimentation. Fourth section will explain the results and its validation. Fifth section will explain the conclusion.

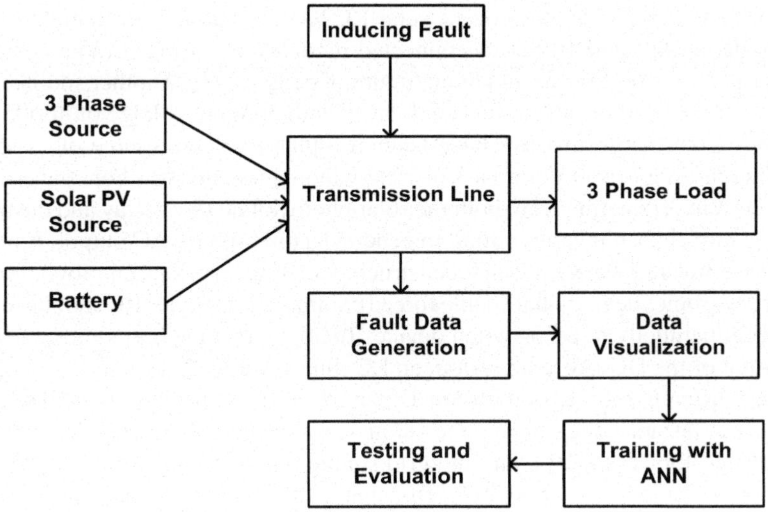

Fig. 5.1 Block diagram of the proposed methodology

5.2 Proposed Methodology

The block diagram of the methodology is shown in Fig. 5.1. Three different sources have been used to feed the transmission line. The output of the solar PV source, battery, and 3 phase AC source is given to the transmission line through inverter and filter. The transmission line is connected to the load. The symmetrical and unsymmetrical faults are created with the help of fault creation block in MATLAB. The generated fault data is then given to the neural network as an input to the neural network. Some fault data samples are used for the testing of the network.

5.3 Methodology Implementation

The methodology is implemented in three steps

1. Fault data generation with the help of MATLAB SIMULINK.
2. Training and testing of the neural network.
3. Validation through results.

5.3.1 Fault Data Generation

The Simulink model is created in MATLAB as shown. The microgrid model is built with three different sources. A renewable energy source is taken which is the Solar source, for which PV array Kyocera Solar KC200GT is used in the simulation with 40 parallel strings and 16 Series-connected modules per string. Irradiance applied to solar panels is 600 W/m^2 and temperature of cells, 25 °C. Another source used is a battery pack which is a Lithium-ion battery with nominal voltage of 460 V. Third source is from the grid, so we have taken a 3-phase voltage source of 11 kV and further reduced it to 460 V with a 3-phase 2 winding transformer. For converting the direct current originating from both the solar photovoltaic (PV) array and the battery, we have modeled an inverter circuit to generate a three-phase AC output. Employed is a 3-level Inverter, where this unit constructs a three-level bridge employing specific power electronics devices that enable forced commutation. Series RC snubber circuits are linked parallelly to each switch device. To counteract the harmonics existing in the output of the DC–AC conversion, an LC filter is utilized.

The 3 different sources outputs are displayed on the scope for a particular interval of time. For instance, the solar source output is displayed for 0 to 1/3 s, battery output is displayed for 1/3 to 2/3 s and the grid source output is displayed for 2/3 to 1 s, considering 10 s of total run time. To achieve all this three-phase circuit breaker block is used after each source, when the external switching time mode is selected, a Simulink logical signal is used to control the breaker operation. For VI measurement, ideal three-phase voltage and current measurement block is used.

Subsequently, the simulation involves the utilization of a Three-phase PI Section Line model, featuring a transmission line spanning 150 km. This model encapsulates a well-balanced three-phase transmission line configuration, where its parameters are condensed into a single PI section. In contrast to the Distributed Parameters Line model, where resistance, inductance, and capacitance are uniformly distributed, the Three-Phase PI Section Line block efficiently groups these parameters into a singular PI section. To induce faults in the transmission line, a Three-phase fault block is implemented, facilitating the introduction of short-circuit faults between any phase and the ground.

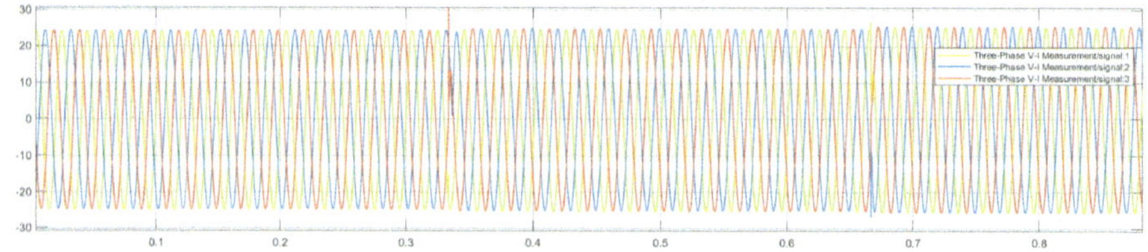

Fig. 5.2 Output waveforms of the microgrid under normal conditions

When the external switching time mode is selected, a Simulink logical signal is used to control the fault operation. Whenever the output is to be measured the Ideal three-phase voltage and current measurement block is used and waveforms as observed on the scope. At the end of the simulation a resistive load is added.

The construction of fault and data generating models was facilitated through the utilization of Simulink. The data generation process was orchestrated via the model depicted in Fig. 5.1. Within this setup, the fault generating block was employed to emulate various types of faults. This encompassed the simulation of a comprehensive array of line faults, namely, RG, BG, YG, RY, YB, RB, RYG, YBG, RYG, RYB, and RYBG. In total, the analysis encompasses 11 distinct fault scenarios that necessitate identification and classification. Notably, when any of these 11 faults manifest, the output signal registers a value of 1 to signify the occurrence of the respective fault during the fault period. Conversely, in the absence of a fault event, the output signal remains at 0.

5.3.2 SCGB-Based Training of ANN

For the training of the Artificial Neural Network (ANN), the Neural Net Pattern Recognition technique was employed. This program provides step-by-step guidance for utilizing a two-layer feed-forward pattern net network, featuring sigmoid output neurons, to address classification challenges in pattern recognition. The ANN is trained through the integrated training algorithm of the Neural Net Pattern Recognition tool. The training process utilizes the scaled conjugate gradient back-propagation method, which involves the adjustment of weight and bias variables. The application of this methodology is realized through the implementation of the "trainscg" function within the Neural Net Pattern Recognition tool.

The conjugate gradient methods, particularly trainscg, appear to perform well across a wide range of tasks, particularly for networks with several weights. On function approximation tasks (faster for bigger networks), the SCG (Scaled conjugate gradient) approach is almost as quick as the LM (Levenberg Marquardt) algorithm, and on pattern recognition challenges, it is almost as fast as trainrp. When the error is lowered, its performance does not degrade as quickly as trainrp's does. The conjugate gradient methods only use a little amount of memory. SCG avoids a time-consuming line search every learning iteration by adopting a step size scaling mechanism, making it faster than other second-order algorithms. Some of the references about scaled conjugate gradient are given here for references (Mishra et al. 2015; Dubois et al. 2008; Khadse et al. 2015, 2016).

In the provided Fig. 5.2, which illustrates the RG fault waveform, the R-phase grounding results in the yellow waveform's amplitude being zero throughout the fault duration. Similarly, the YG and BG faults exhibit analogous behavior, where the phases are grounded, leading to a linear waveform for the respective phases during the fault interval. In the context of the RY fault, both the R and Y phases experience a short circuit, inducing synchronous oscillations and subsequent voltage reduction during the fault period. Corresponding patterns are observed for the YB and RB faults.

During an RYG fault, the R and Y phases are grounded, resulting in a constant waveform for these phases, while the B phase waveform remains discernible in the voltage waveform. This principle extends to YBG and RBG faults as well. In the instance of an RYB fault, all three phases R, Y, and B are subjected to a short circuit, inducing complete voltage attenuation during the fault interval. This phenomenon is replicated in the RYBG fault scenario, which also yields negligible voltage output.

- The train curve cross entropy attains a stable value after almost 10 epochs and stabilizes at a constant value for 315 epochs which is shown in Fig. 5.3.
- The cross entropy for validation and test also traces the similar constant value for all epochs.
- The Best Validation Performance is 0.27213 at 319 epochs.

Training Results

Training finished: Met validation criterion

Training Progress

Unit	Initial Value	Stopped Value	Target Value
Epoch	0	325	1000
Elapsed Time	-	00:01:06	-
Performance	0.499	0.275	0
Gradient	0.336	1.31e-05	1e-06
Validation Checks	0	6	6

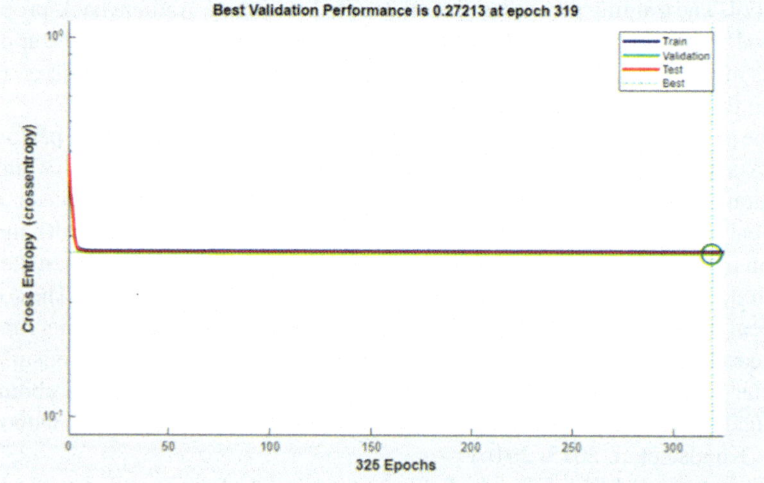

Fig. 5.3 Training, testing, and validation performance

5.4 Results of the Implemented Methodology

1. Phase to Ground: Phase-to-ground faults involve a single phase being grounded, resulting in a generated fault. During the fault duration, the output is 1; once resolved, it reverts to 0. The fault model block was utilized to generate distinct faults for all three phases. Figures 5.4, 5.5 and 5.6 show the acquired results.
2. Phase to Phase: Phase-to-Phase faults occur when both phases are shorted, leading to a fault occurrence. Throughout the fault period, the output registers as 1, transitioning back to 0 upon resolution. The fault model block was employed to create discrete faults for each of the three phases. Figures 5.7 and 5.8 show the acquired results.

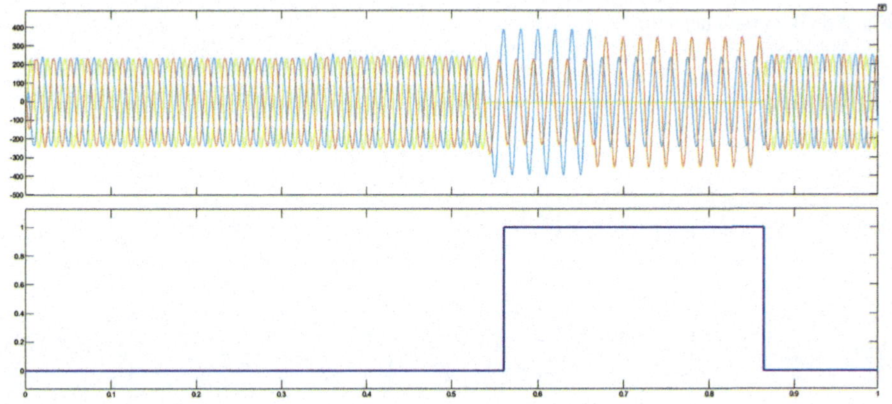

Fig. 5.4 R–G fault detection

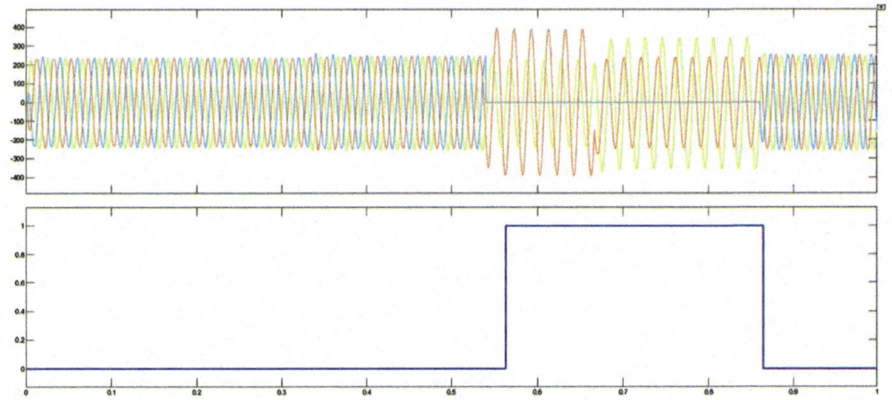

Fig. 5.5 Y–G fault detection

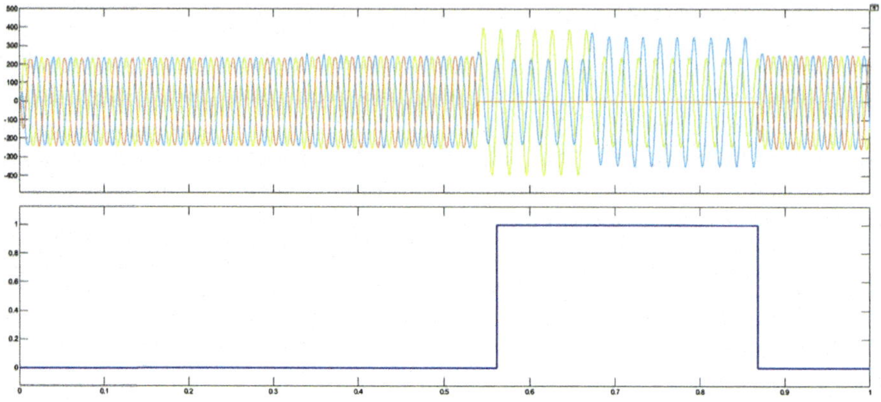

Fig. 5.6 B–G fault detection

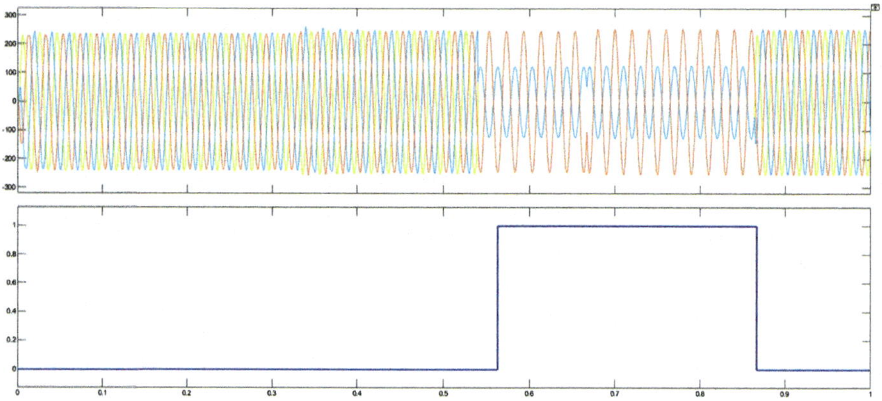

Fig. 5.7 R–Y fault detection

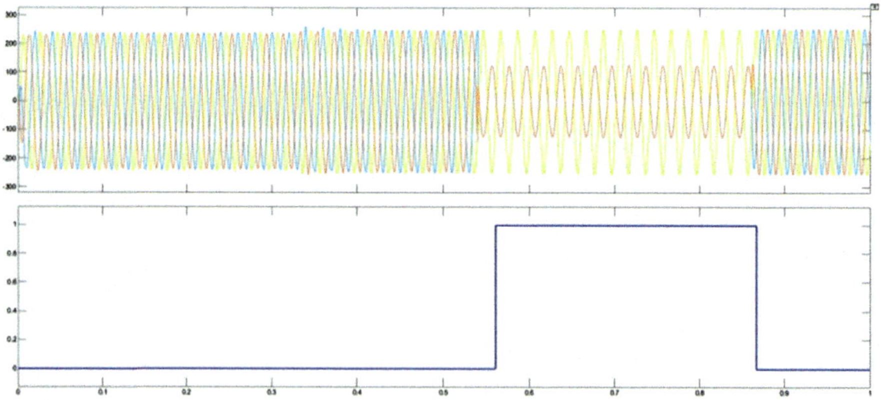

Fig. 5.8 Y–B fault detection

3. Double Phase to Ground: Double Phase-to-Ground faults manifest when both phases are shorted with the ground, resulting in a fault event. The output maintains a value of 1 throughout the fault duration, reverting to 0 once the fault is resolved. To simulate these distinct faults across the three phases, the fault model block was utilized. Figure 5.9 shows the acquired result.
4. Three Phase: Three-phase faults occur when all three phases are shorted among themselves, leading to a fault event. The output signal remains at a value of 1 throughout the fault duration and returns to 0 upon fault resolution. The fault model block was employed to simulate these specific faults across all three phases. It is similar to the fault shown in Fig. 5.10.
5. Three Phase to Ground: Three-phase-to-Ground faults arise when all three phases are shorted with the ground, resulting in a fault event. The output signal maintains a value of 1 throughout the fault duration, transitioning back to 0 upon

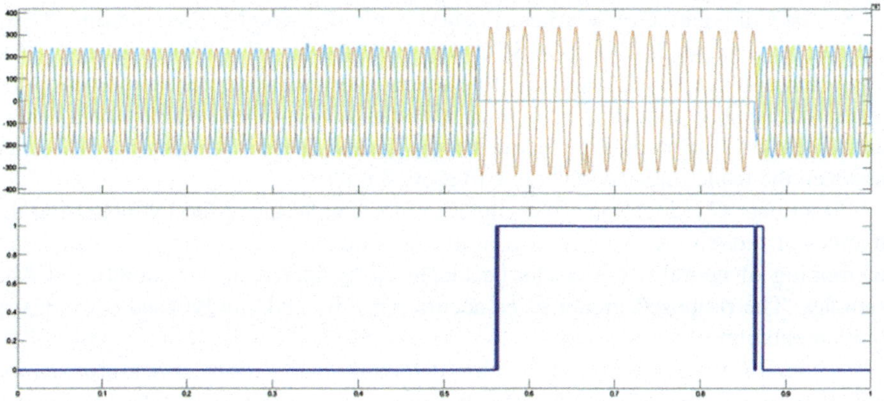

Fig. 5.9 R–Y–G fault detection

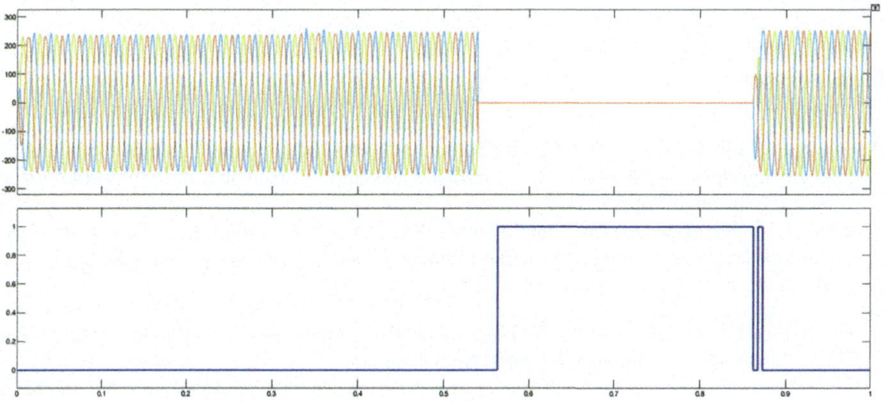

Fig. 5.10 R–Y–B fault detection

the resolution of the fault. To replicate these particular fault scenarios across all three phases, the fault model block was utilized. Figure 5.10 shows the acquired results.

5.5 Conclusion

This study delves into the application of artificial neural networks for the detection and classification of faults within a small transmission line of the microgrid. The research presents outcomes pertaining to line to ground, line to line, double line to ground, three phase fault, and three phase to ground fault. The architecture employed for the neural network is based on back-propagation. The simulation findings illustrate that the neural networks have achieved commendable performance and hold practical feasibility for implementation.

Scaled Conjugate Gradient-based neural network is used for the training. There are no user-dependent settings in SCG whose values are critical to its success. This algorithm avoids a time-consuming line search every learning iteration by adopting a step size scaling mechanism, making it faster than competing algorithms. There are some minor errors during classification of faults. There is a negligible difference between the triple line and triple line to ground fault.

If there are classification errors, the accuracy can be increased by increasing the number of hidden layers and selecting the best number of neurons. The efficiency of training of neural network can be increased by increasing the number of data samples. The proposed model gives accurate results and can be used for industry implementation.

Further to this project fault location detection can be done. Back-propagation neural networks perform well when trained with a large training data set, which is simple to build in MATLAB, and hence back-propagation neural networks were chosen for this project.

References

Dubois A, Dubois T, Boorman C, Connor S, Poole (2008) An implementation of the conjugate gradient algorithm on FPGAs. In: 16th International symposium on field-programmable custom computing machines, pp 296–297

Esmaeilbeigi S, Karegar HK (2020) Intelligent fault detection and location scheme for low voltage microgrids based on recurrent and radial basis function neural networks. In: 2020 28th Iranian conference on electrical engineering (ICEE), Tabriz, Iran

Grcić I, Pandžić H (2023) Artificial neural network for high-impedance-fault detection in DC microgrids. In: 2023 IEEE PES conference on innovative smart grid technologies—Middle East (ISGT Middle East), Abu Dhabi, United Arab Emirates

Khadke P, Patne N, Singh A et al (2016) A soft computing scheme incorporating ANN and MOV energy in fault detection, classification and distance estimation of EHV transmission line with FSC. SpringerPlus 1834

Khadse CB, Chaudhari MA, Borghate VB (2016) Conjugate gradient back-propagation based artificial neural network for real time power quality assessment. Int J Electr Power Energy Syst 82:197–206

Khadse C, Patharkar AA, Chaudhari BS (2021) Electromagnetic field and artificial intelligence based fault detection and classification system for the transmission lines in smart grid. Energy Sources, Part A: Recover, Util, Environ Eff. https://doi.org/10.1080/15567036.2021.1948637

Khadse CB, Chaudhari MA, Borghate VB (2017) Electromagnetic compatibility estimator using scaled conjugate gradient backpropagation based artificial neural network. IEEE Trans Industr Inf 13(3):1036–1045

Khadse CB, Chaudhari MA, Borghate VB (2015) Comparison between three back-propagation algorithms for power quality monitoring. In: 2015 Annual IEEE India conference (INDICON), New Delhi, India, pp 1–5. https://doi.org/10.1109/INDICON.2015.7443766

Khadse CB, Chaudhari MA, Borghate VB (2017) Comparison of seven backpropagation algorithms for three phase power quality assessment. In: TENCON 2017—2017 IEEE region 10 conference, Penang, Malaysia

Lunden J, Koivunen V (2007) Scaled conjugate gradient method for radar pulse modulation estimation. In: 2007 IEEE international conference on acoustics, speech and signal processing—ICASSP '07, Honolulu, HI, USA, pp II-297–II-300

Mishra S, Prusty R, Hota PK (2015) Analysis of Levenberg–Marquardt and scaled conjugate gradient training algorithms for artificial neural network based LS and MMSE estimated channel equalizers. In: 2015 International conference on man and machine interfacing, pp 1–7

Ogar VN, Hussain S, Gamage KAA (2023) The use of artificial neural network for low latency of fault detection and localisation in transmission line. Heliyon 9(2)

Pandey A, Gadekar PS, Khadse CB (2021) Artificial neural network based fault detection system for 11 kV transmission line. In: 2021 International conference on advances in electrical, computing, communication and sustainable technologies (ICAECT), Bhilai, India, pp 1–4

Singh AR, Lei D, Kumar A, Singh R, Meena NK (2020) Microgrid system. In: Ray P, Biswal M (eds) Microgrid: operation, control, monitoring and protection (Lecture notes in electrical engineering), vol 625. Springer, Singapore

Sivasankari K, Thanushkodi K, Kalaivanan K (2013) Automated seizure detection using multilayer feed forward network trained using scaled conjugate gradient method. In: 2013 International conference on current trends in engineering and technology (ICCTET), Coimbatore, India

Zubir NSA et al (2017) Pattern classifier of chemical compounds in different qualities of agarwood oil parameter using scale conjugate gradient algorithm in MLP. In: 2017 IEEE 13th international colloquium on signal processing & its applications (CSPA), Penang, Malaysia, pp 18–22

Chapter 6
IoT-Based Intelligent Home Automation System Using IFTTT with Google Assistant

Mohan Bansal, Rajeev Sharma, Ayush Yadav, and Krishna

Abstract Intelligent home automation system (IHAS) with Internet of Things (IoT) can improve the quality of life for individuals by automating daily tasks, such as controlling the temperature of a room and turning off lights when not needed. Additionally, IoT devices can also provide real-time monitoring and security, allowing users to keep track of their home even when they're not there. Furthermore, the use of IoT in intelligent home automation can also lead to energy savings by automatically turning off devices when they're not in use and optimizing the usage of energy-efficient devices. It can also reduce the waste of energy by automatically turning off appliances that have been left on. In this chapter, an IoT-based IHAS prototype is designed using If This Then That (IFTTT) and Google Assistant. The system allows users to control and automate various devices in their home through voice commands in multiple languages given to Google Assistant or through Adafruit IO from anywhere on earth. The system integrates various alternating current (AC) and direct current (DC) devices such as electric bulb, electric socket, DC motor, and light emitting diodes (LEd) lights. This system allows them to communicate with each other to perform tasks based on user preferences. The system operates on the logic created in IFTTT, where the user can set up conditions or triggers for a specific action to take place. This allows for a personalized and flexible home automation experience for the user. The integration with Google Assistant further enhances the ease of use, allowing the user to control their smart home devices through simple voice commands in any language. The multilingual voice control system may enhance the system robustness and performance.

M. Bansal (✉)
Electronics and Communication Engineering, Indian Institute of Information Technology, IIIT Sonepat, Khewra, Haryana, India
e-mail: bmohan@iiitsonepat.ac.in

R. Sharma · A. Yadav · Krishna
VIT-AP University, G-30, Inavolu, Beside AP Secretariat Amaravati, Amaravati, Andhra Pradesh 522237, India
e-mail: rajeev.sharma@vitap.ac.in

A. Yadav
e-mail: ayush.21bce8020@vitapstudent.ac.in

Krishna
e-mail: krishna.21bce8114@vitapstudent.ac.in

C. B. Khadse et al. (eds.), *Intelligent Methods in Electrical Power Systems*, Engineering Optimization: Methods and Applications, https://doi.org/10.1007/978-981-97-5718-3_6

Keywords Intelligent home automation system · IoT · Adafruit IO · Google Assistant · IFTTT

6.1 Introduction

Internet of Things (IoT)-based systems make use of the diversity of interconnection and communication among various heterogeneous devices/sensors to execute a specified task (Li et al. 2015) from remote locations in collaboration. IoT devices can be connected to each other by the Internet, allowing them to share data and interact with each other without human intervention. IoT devices can be controlled and monitored remotely, providing increased efficiency, convenience, and insights into how we live and work. The IoT is a rapidly growing field and is expected to have a significant impact on our lives in the years to come (Munirathinam 2020).

The trend of home automation is on the rise in India due to the low cost and accessibility of devices. The home automation systems are generally intended to lower energy consumption, boost the efficiency of household appliances, and enhance comfort and security features (Reinisch et al. 2010). Automatic electrical and lighting equipment can aid in the prevention of accidents including short circuits, fires, and energy waste. People frequently forget to switch off their appliances while using conventional methods, which can cause confusion and conflict when they must go back home to do so. Such occurrences may be avoided by modern technologies. The advantages of a home automation system are shown in Fig. 6.1.

The IHAS, accessible through mobile apps offer the convenience of remotely controlling and monitoring various aspects of a home's energy usage, thereby improving

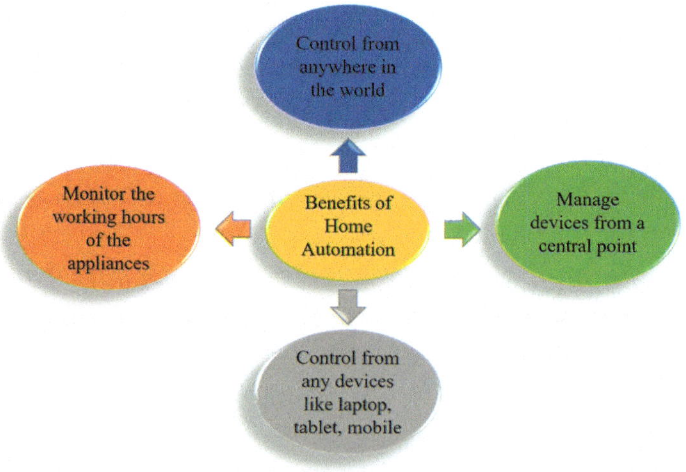

Fig. 6.1 Benefits of intelligent home automation system

energy efficiency. With just a few taps on a mobile device, homeowners can turn off lights, and even power down appliances from anywhere in the world. This level of control enables individuals to actively manage energy consumption, ensuring that lights and devices are not left on unnecessarily. Additionally, IHAS can provide real-time energy usage data, allowing homeowners to identify areas of high energy consumption and make informed decisions to optimize efficiency. By harnessing the power of mobile apps for smart home automation, individuals have the tools at their fingertips to reduce energy waste, conserve resources, and contribute to a greener and more sustainable future.

IoT-based home automation systems are designed by many researchers to automate and control various household functions using smart devices, sensors, and cloud-based services (Kung et al. 2018; Abdulraheem et al. 2020; Chaurasia and Jain 2019; Jie et al. 2013). These systems provide homeowners with convenience, energy efficiency, enhanced security, cost-effectiveness, and customization (Luan and Leng 2016; Kim et al. 2017). IoT-based home automation systems by using a combination of hardware and software technologies can help to create a seamless and convenient living experience for homeowners.

The hardware components of an IoT-based home automation system typically include smart devices, sensors, controllers, and a central hub (Dash and Choudekar 2021). Controllers such as Raspberry Pi, Arduino, and NodeMCU boards can be used to connect and control various smart devices and sensors (Hadwan and Reddy 2016). Raspberry Pi, NodeMCU, and Arduino are all popular open-source platforms for building electronics projects, but they differ in their capabilities and intended uses. Raspberry Pi has its own advantages, such as more powerful processing capabilities, built-in multimedia support such as Ethernet, Wi-Fi, and Bluetooth, and the ability to run full-fledged operating systems. Finally, a central hub such as a mobile app or a web-based dashboard is used to control and monitor various devices and sensors remotely.

The software components of an IoT-based home automation system typically include cloud-based services, communication protocols, and automation rules (Venkatesh et al. 2018). Cloud-based services such as Google Cloud, Amazon Web Services, and Microsoft Azure can be used to store and process data from smart devices and sensors. To connect and communicate with various smart devices and sensors, communication protocols such as Zigbee, GSM, RF, Bluetooth, and Wi-Fi can be implemented (Morshed et al. 2015). Finally, automation rules are used to automate various household functions based on predefined conditions, such as time of day, temperature, and user input.

In this chapter, an IoT-based framework is developed to work in the intelligent home automation to reduce energy consumption, and improve comfort and quality of life especially for older people or physically disabled person. Intelligent home technology allows users to control and monitor various devices at home from a single device such as a smartphone or tablet (Wilson et al. 2017). This makes it easier and more convenient to manage home and its systems. Users can monitor and control the energy usage more effectively by using smart devices that are connected to the Internet. For example, user can set lights/fans/AC to turn on/off, even after leaving the home from a remote location with the help of IoT.

The objective is to develop an IHAS that can manage various electric loads using voice commands. By utilizing Raspberry Pi, creating a smart home can be simple and effective. The aim is to provide a user-friendly, reliable, and adaptable home automation system using Android devices. This project enables users to control electric appliances within their homes by speaking voice commands from any location. Raspberry Pi has the added benefit of integrating with this home automation system, allowing users to manage their appliances and electronic systems using their smartphones. The system uses relay circuits and IoT concepts to turn appliances and electrical devices on or off, thereby making their use convenient and efficient.

The study aims to create a portable IHAS that provides convenient monitoring and remote control of home appliances via the Internet, regardless of the location. The study aims to accomplish the following primary goals:

1. To create a cutting-edge prototype IHAS that utilizes Raspberry Pi and an IoT platform.
2. To develop an IoT-based algorithm, which can manage IHAS that allows for seamless integration with various home appliances.
3. To develop an algorithm which can support multilingual voice commands to control the appliances from a remote location.

6.2 Related Prior Work

There has been a significant amount of prior research and development in the field of intelligent or smart home automation systems (Krishna and Lavanya 2017; Gunge and Yalagi 2016; Shabber et al. 2021; Hettiarachchi et al. 2022; Moon and Shin 2022; Jabbar et al. 2018). The existing literature on smart home automation systems focuses on the types of sensors, communication technologies, and control strategies used in these systems (Sharma and Reddy 2022). In general, the home automation systems consist of a central control unit, and various peripheral devices such as sensors, actuators, and relays that are connected to the control unit (Skeledzija et al. 2014).

IHAS offers superior connectivity, scalability, compatibility, user interface, energy efficiency, security, and cost-effectiveness compared to traditional home automation systems. IHAS utilizes wireless technologies like Zigbee, GSM, RF, Wi-Fi, and Bluetooth for seamless communication between devices and central control hubs. They provide intuitive user interfaces through smartphone apps, web portals, or voice assistants for remote control and monitoring. They also prioritize security through data encryption and secure authentication. While IHAS may have higher initial costs, they offer long-term savings and advancements over traditional systems. In the following section, we review related works with their advantages and disadvantages.

6.2.1 *Zigbee-Based IHAS*

Zigbee-based IHAS uses the Zigbee wireless communication protocol to control and automate various household appliances and devices (Gill et al. 2009). Zigbee uses a low-data rate, low-power, and low-cost wireless communication protocol that is specifically designed for IoT applications. Zigbee-based home automation systems consist of a central control unit, which is usually a Zigbee hub (John and Santhosam 2014). The central control unit acts as a bridge between the various peripheral devices and the user's mobile device or computer. It also processes commands from the user and transmits the commands to the peripheral devices.

One advantage of Zigbee-based home automation systems is that they are designed for low-power consumption, which makes them ideal for IoT applications and helps to extend the battery life of the peripheral devices (Baviskar et al. 2015). Another advantage is that Zigbee supports mesh networking, which allows multiple devices to form a self-healing network that can automatically route signals around any failures or obstacles in the network.

However, there are also some disadvantages of Zigbee-based home automation systems. One disadvantage is that Zigbee has a limited range compared to other wireless communication protocols, so the user may need to place additional Zigbee hubs in the home to ensure full coverage (Khan et al. 2016). Another disadvantage is that Zigbee is a proprietary protocol, so users have limited devices that are specifically designed to work with the Zigbee protocol.

6.2.2 *GSM-Based IHAS*

Global system for mobile communications (GSM)-based home automation systems use the GSM network to control and automate various household appliances and devices (Tamakloe and Kommey 2022). GSM-based home automation systems consist of a central control unit, which is usually a micro-controller or a single-board computer (Jivani 2014). The central control unit is responsible for receiving commands from the user's mobile phone and processing these commands to control and automate the various devices and appliances in the home. Users can easily control and automate their devices by simply sending a text message or making a phone call to the central control unit, which then executes the corresponding action. This seamless communication ensures that users can conveniently manage their smart homes with ease.

One advantage of GSM-based home automation systems is that they can be controlled from anywhere with a mobile phone, as long as there is a GSM network available (Vaidya and Vishwakarma 2018). This means that users can control and automate their homes even when they are not physically present. Another advantage of GSM-based home automation systems is that they are relatively simple and inexpensive to set up and maintain, compared to other smart home automation systems.

However, there are also some disadvantages to GSM-based home automation systems (Vaidya and Vishwakarma 2018). One disadvantage is that the system is dependent on the availability and quality of the GSM network, so if there is a problem with the network, the system may not be able to function properly. Another disadvantage is that the system may not be as secure as other smart home automation systems, as text messages and phone calls can be intercepted or redirected by malicious actors.

6.2.3 RF-Based IHAS

Radio frequency (RF)-based home automation systems use RF signals to control and automate various household appliances and devices (Gu 2006). RF-based home automation systems consist of a central control unit, which is usually a microcontroller or a single-board computer. Users can easily control and automate their devices by simply sending a command using the remote control or mobile device to the central control unit via RF signals (Jiang et al. 2004).

One advantage of RF-based home automation systems is that they are relatively simple and inexpensive to set up and maintain, compared to other smart home automation systems (Mittal et al. 2015). Another advantage is that RF signals can be transmitted over a relatively long distance, so the user can control and automate their home from a distance.

However, there are also some disadvantages to RF-based home automation systems. One disadvantage is that the system is susceptible to interference from other RF signals in the environment, which can result in incorrect or delayed responses (Hoque and Davidson 2019). Another disadvantage is that the system may not be as secure as other smart home automation systems, as RF signals can be intercepted or redirected by malicious actors.

6.2.4 Wi-Fi-Based IHAS

Wi-Fi-based home automation systems use Wi-Fi wireless communication to control and automate various household appliances and devices (ElShafee and Hamed 2012). Wi-Fi-based home automation systems consist of a central control unit, which is usually a smart hub. Users can easily control and automate their devices by simply sending a command using mobile app or web interface to the central control unit via Wi-Fi. The central control unit processes the command and performs the corresponding action.

One advantage of Wi-Fi-based home automation systems is that they offer high-speed and reliable wireless communication, which allows for fast and responsive control and automation of the devices and appliances in the home

(Kodali et al. 2016). Another advantage is that Wi-Fi is widely available and compatible with a wide range of devices, so users can choose from a large selection of compatible devices and accessories.

However, there are also some disadvantages to Wi-Fi-based home automation systems. One disadvantage is that Wi-Fi-based home automation systems require a strong and stable Wi-Fi connection, so if there are any issues with the Wi-Fi network, the system may not function properly. Another disadvantage is that Wi-Fi-based home automation systems can be vulnerable to hacking and cyber attacks, as they are connected to the Internet and can be accessed from remote locations (Ali 2022).

6.2.5 Bluetooth-Based IHAS

Bluetooth-based home automation systems use Bluetooth wireless communication to control and automate various household appliances and devices (Sriskanthan et al. 2002). These systems consist of a central control unit, which is usually a smart hub or a smartphone. When the user wants to control or automate a device or appliance, they send a command using a mobile app or other software on their smartphone, which is then transmitted to the central control unit via Bluetooth. The central control unit processes the command and performs the corresponding action.

However, there are also some disadvantages to intelligent home automation systems using Bluetooth devices. One disadvantage is that Bluetooth has a limited range compared to other wireless communication protocols, so the user may need to be within close proximity to the central control unit to control and automate their devices (Asadullah and Raza 2016). Another disadvantage is that Bluetooth-enabled devices may be vulnerable to hacking and cyber attacks, as they can be accessed from remote locations.

6.3 Proposed Methodology

The proposed system IHAS is implemented to design intelligent home automation to control the home appliances from a remote location in multiple languages. The primary goal of this system is to create a seamless and convenient living experience for homeowners by integrating various smart devices and sensors.

The IHAS is based on a Raspberry Pi board that will act as a central hub to connect and control various smart devices and sensors. The system will use IFTTT (If This Then That) to automate tasks and create applets that can be triggered by various events, such as voice commands, sensor data, and other inputs. Google Assistant is used as the primary voice interface for the system, allowing users to control various devices and functions with voice commands. The Adafruit IO platform is utilized to collect and store data from sensors and smart devices, which will be used to trigger automated actions or provide real-time information about the state of the home. The system consists of several components, which are discussed in Sect. 6.3.1.

6.3.1 Required Resources

In this implementation, we have utilized a range of technologies and devices to create a highly efficient prototype of IHAS. Our work presents a novel approach to intelligent home automation by combining Google Assistant, IFTTT, Adafruit IO, Raspberry Pi 3B board, and a 4-channel relay module to enable seamless control of up to four appliances simultaneously. It explores the possibilities of leveraging Google Assistant's voice control, IFTTT's flexible automation capabilities, and Adafruit IoT platform to create a comprehensive and user-friendly home automation solution. The system automates different home functions like lighting, security, and entertainment, and can be conveniently controlled using voice commands through Google Assistant. A mobile app is also created to control the system's devices and functions from a remote location. The system is designed to address the daily issues people face, with a focus on affordability and convenience.

Google Assistant: Google Assistant employs cutting-edge natural language processing and machine learning algorithms to accurately comprehend and address user inquiries. Its multi-functional abilities enable it to carry out a diverse set of tasks, including scheduling reminders, providing information, playing music, managing smart home devices, and many others (Akinbi and Berry 2020). Google Assistant also supports multi-lingual capabilities, allowing users to interact with it in multiple languages. Additionally, it can recognize multiple users, allowing each person to access their own personalized information, such as their reminders and calendar events. Overall, Google Assistant provides a convenient and easy-to-use virtual assistant that can help simplify and streamline a wide range of tasks and activities. With its natural language processing capabilities and integration with a wide range of services, Google Assistant makes it easy for users to get things done hands-free. Google Assistant can be set up to allow for an intelligent home automation system by voice-control of various smart devices in a home (Gupta 2018).

IFTTT: If This Then That (IFTTT) is a web-based platform that allows users to create connections between different services and devices, known as "applets" (Ovadia 2014). These applets allow users to automate simple tasks and create workflows that can be triggered by events and actions. IFTTT allows user to automate various tasks by creating simple "recipes" based on triggers and actions (Ur et al. 2016). A trigger is an event that initiates a workflow. An action is something that occurs as a result of a trigger. When used in conjunction with Google Assistant and Adafruit IO, IFTTT can be used to create powerful home automation solutions. Once the recipe is created for the home automation system, it can be triggered by using Google Assistant. For example, users can say "Hey Google, turn on the light," which would then trigger the IFTTT recipe and take an action on Adafruit IO dasboard to turn on the light.

Adafruit IO: Adafruit IO is a cloud-based platform for Internet of Things (IoT) that enables users to collect, store, and display data from connected devices (Rubell 2018). It uses feed for collection of data points that are sent to Adafruit IO by a connected device. Users can create multiple feeds to store different types of data, such as temperature readings, sensor data, and more. It also provides a RESTful API that enables users to interact with their data programmatically. They can use the API to retrieve data, send data to feeds, and perform other operations. Adafruit IO integrates with a range of platforms and services, including home automation systems, mobile devices, and more. This makes it easier for users to integrate their projects with other tools and services.

Adafruit IO is an online application to create a dashboard as shown in Fig. 6.2. Users can control the home appliances manually or based on the voice command given to Google Assistant with the help of Adafruit IO. IFTTT is used to attach Adafruit IO with Google Assistant.

Raspberry Pi: The Raspberry Pi boards are based on a System on a Chip (SoC) design, which integrates all the necessary components, such as a CPU, GPU, RAM, and other peripherals, onto a single-circuit board (Richardson and Wallace 2012). They are available in several different models, with varying specifications, such as the number of CPU cores, the amount of RAM, and the size of the General Purpose Input/Output (GPIO) header.

The versatility of Raspberry Pi boards is evident in the diverse array of projects and applications, such as media centers, gaming consoles, home automation systems, robots, and much more. They are also popular in educational settings, where they are used to teach children about computer science and programming. Overall, Raspberry Pi is a powerful and versatile single-board computer that provides a low-cost and

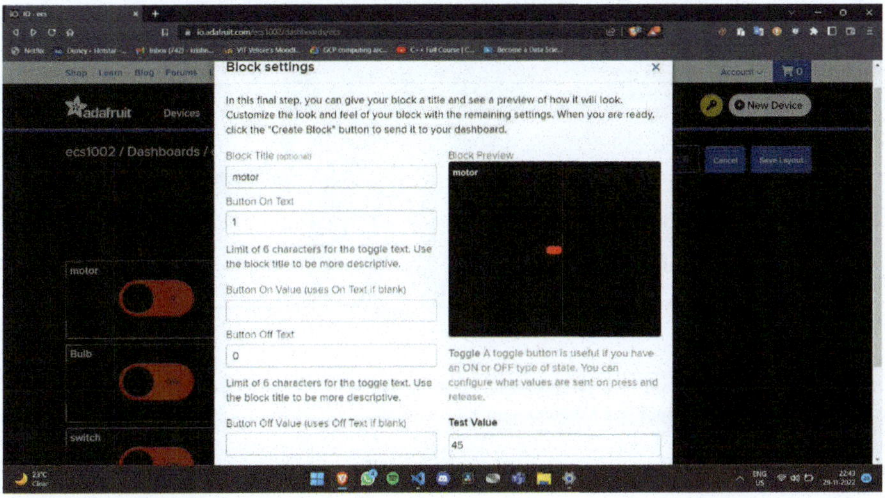

Fig. 6.2 Dashboard creation in Adafruit

accessible platform for a wide range of projects and applications. With its processing power, input and output options, connectivity, and community, Raspberry Pi is an excellent choice for anyone looking to explore the world of single-board computing.

The work utilizes the Raspberry Pi 3 Model B (3B) (Pi 2015) and a labeled diagram of the board is shown in Fig. 6.3. However, to provide a more comprehensive understanding, a detailed description of the board is presented below.

CPU: Raspberry Pi 3B boasts a 1.2GHz Broadcom BCM2837B0 processor with four 64-bit ARM Cortex-A53 cores, providing high-performance computing. Additionally, it has 1 GB of RAM to support multitasking and memory-intensive tasks.

Storages and ports: Raspberry Pi 3B has a microSD card slot for storage. It also has four USB 2.0 ports, a full-sized HDMI port, an Ethernet port, a 3.5mm audio jack, and a camera interface as shown in Fig. 6.3.

Wireless connectivity: Raspberry Pi 3B has built-in Bluetooth 4.1 and Wi-Fi (802.11b/g/n). This makes it easy to connect the board to the Internet and other devices without the need for additional hardware.

GPIO: Raspberry Pi 3B has a 40-pin GPIO header that can be used to connect a variety of sensors, displays, and other devices.

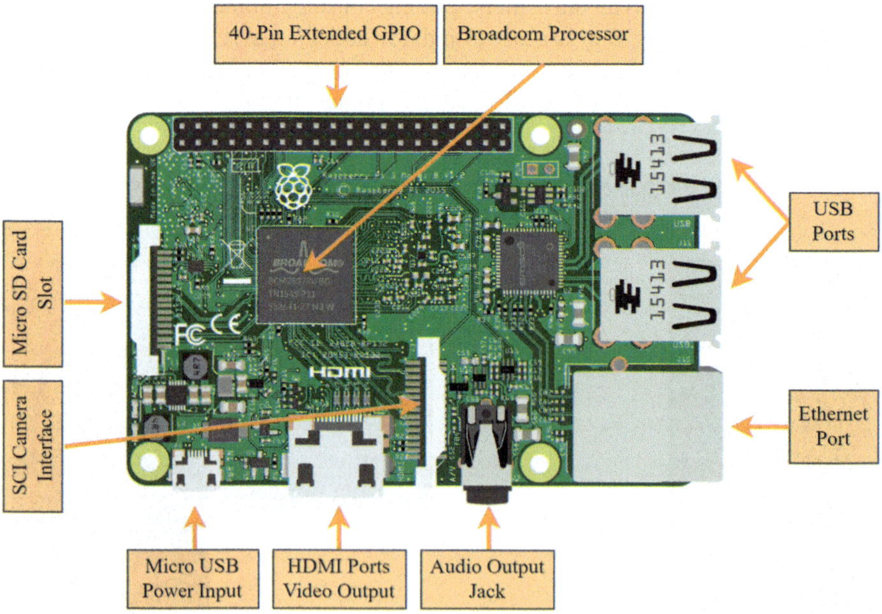

Fig. 6.3 Raspberry Pi 3B board

Operating system: The versatility of Raspberry Pi 3B extends to its ability to support a diverse range of operating systems, such as Raspbian (based on Debian Linux), Ubuntu, and even Windows.

Power: Raspberry Pi 3B can be powered by a 5V micro USB power supply. It also has a power-over-ethernet (PoE) header that allows it to be powered by an ethernet cable with a compatible PoE switch.

Four-Channel Relay Module: A 4-channel relay module is an electronic device that provides a convenient way to control electrical appliances and devices using a micro-controller, such as Arduino or Raspberry Pi Yunardi et al. (2022). A relay provides a convenient way to control electrical devices, such as motors, lights, and pumps, without requiring a direct connection between the micro-controller and the device. It is essentially an electrically operated switch that can be toggled to turn the device on or off. The schematic diagram of the 4-channel relay module with the connection of Raspberry Pi and appliances are shown in Fig. 6.4.

The 4-channel relay module typically consists of a small circuit board with four relays, an opto-isolator for protection, and a set of connectors for controlling the relays and connecting to the micro-controller. Each relay channel can be independently controlled by the micro-controller, allowing multiple devices to be switched on and off at the same time (Yadav et al. 2018).

The relay module is controlled by sending a digital signal from the micro-controller to the relay module. When the signal is received, the relay switches on or off, depending on the state of the signal. This allows the micro-controller to control

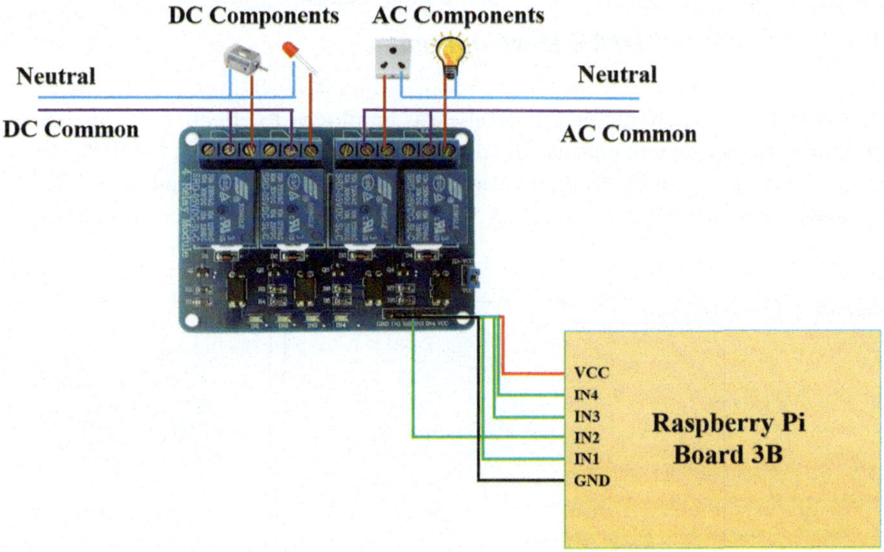

Fig. 6.4 Four-channel relay module with connections

the electrical devices with a simple digital output, without the need for complex circuitry or high voltage protection.

6.3.2 *Load Utilization*

Intelligent home automation systems (IHAS) have the capability to manage an extensive range of appliances and devices throughout the household, including both direct current (DC) and alternating current (AC) appliances, as well as home entertainment systems and security cameras. The following are some typical examples of the load utilization that can be incorporated into IHAS.

DC load: DC appliances are electrical devices that operate using DC power instead of AC power. DC appliances are commonly used in applications where DC power sources are available, such as in battery-operated devices, solar-powered systems, and automotive systems. A DC motor and an array of LED lights are used in this work as DC loads. The images of the DC motor and LEDs are depicted in Fig. 6.5.

AC load: AC appliances are electrical devices that operate using AC power, which is the most commonly available form of power in households and businesses around the world. AC appliances typically use a standard voltage of 220 volts and a frequency of 50 Hz. An electric bulb and an electric socket are used in this work as AC loads. The images of the electric bulb and electric socket are depicted in Fig. 6.6.

6.3.3 *IoT-Based IHAS Development*

In this study, a systematic methodology was employed to effectively structure and execute various research phases. The design and implementation of the IHAS and its hardware prototype are executed with meticulous attention to detail. The selection of components and their integration were thoughtfully considered to achieve the

Fig. 6.5 Utilized DC loads

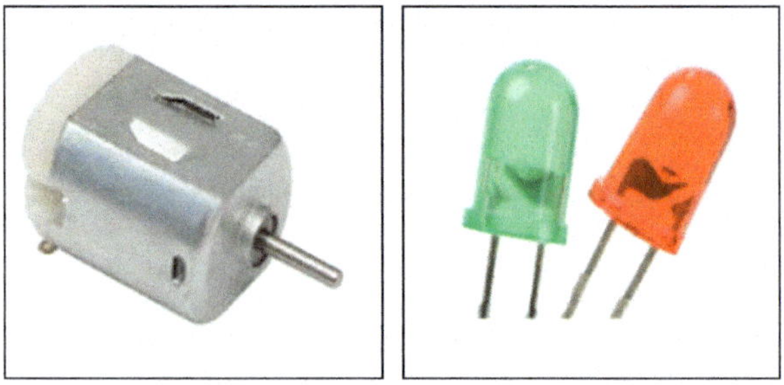

(a) DC motor (b) LEDs

(a) Electric bulb (b) Electric socket

Fig. 6.6 Utilized AC loads

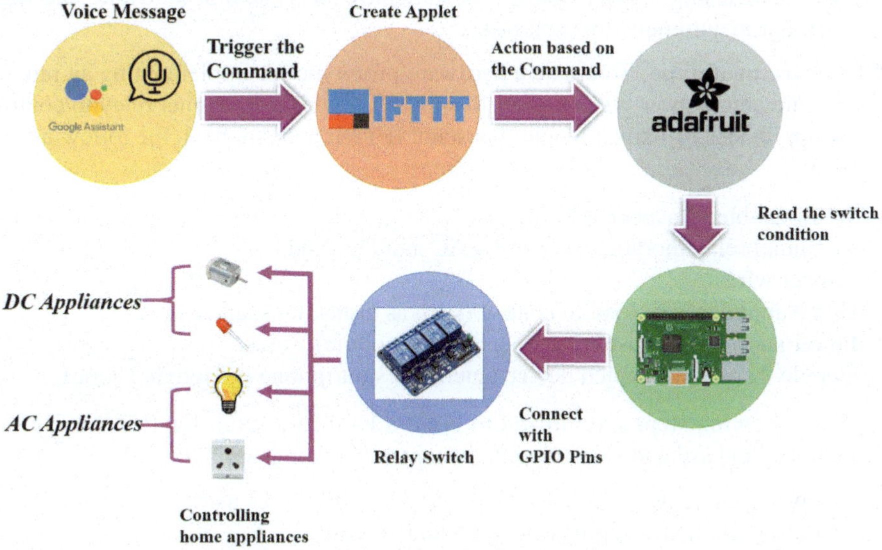

Fig. 6.7 A flow diagram of the proposed system

desired design objectives. The flow diagram of the proposed system is represented in Fig. 6.7. The conceptual framework of the study is illustrated in the following subsection, which depicts the step-by-step procedure of the entire process.

Creating connection of IFTTT with Google Assistant and Adafruit: This methodology provides a framework for building an IHAS using Adafruit, IFTTT, and Google Assistant. By using these technologies, users can create a highly customizable and flexible system that can be easily operated from a remote location by a smartphone. The following are the steps to achieve these connections:

1. Create an account on Adafruit IO to link with IFTTT.
2. Create an IFTTT account, which allows to create applets and link different services and devices together.
3. Link Adafruit IO with IFTTT using the Adafruit service page on IFTTT. This allows IFTTT to access Adafruit dashboard and data.
4. Connect Google Assistant with IFTTT by enabling the Google Assistant service on IFTTT.
5. Create IFTTT applets that link Adafruit IO, Google Assistant, and other desired services. For example, creating an applet that responds to the command "Hey Google, turn on the light" by activating the light in our home.
6. Test the IHAS by executing different commands through Google Assistant and observing the responses of the virtual devices on the Adafruit IO dashboard.
7. Once the testing is successful, we can deploy the system in hardware and start using it to automate different tasks.

Hardware prototype: To create a hardware prototype for implementing an intelligent home automation system using Raspberry Pi with a 4-channel relay to control home appliances by using Google Assistant, IFTTT, and Adafruit, the following are needed:

- Wi-Fi-enabled Raspberry Pi 3B
- 4-channel relay module to control 4 AC and DC loads
- Jumper wires
- Home appliances or load to control (such as lights, fans, and motor)
- Power supply (for Raspberry Pi and relay module)
- Google Assistant-enabled device (such as a smartphone or Google Home).

The following steps are utilized to build a hardware setup for IHAS using the Raspberry Pi board and relay module to control the home appliances.

1. Connect the relay module to Raspberry Pi's GPIO pins using jumper wires.
2. Connect the home appliances to the relay module.
3. Connect the Google Assistant service and Adafruit IO service to the IFTTT account and set up the trigger and action for an applet.
4. Obtain an Adafruit IO account and create feeds for the different appliances that we want to control. Generate an Adafruit IO key and username that will be used by Raspberry Pi to connect to Adafruit IO.
5. Install the necessary software libraries on Raspberry Pi to control the relay module and integrate with Google Assistant, IFTTT, and Adafruit IO.
6. Write code to control the relay module using the GPIO pins, and to interpret voice commands from Google Assistant.
7. Use the Google Assistant software library to recognize voice commands and trigger the appropriate relay control function. Use the Adafruit IO library to send data to the appropriate feed based on the command received.
8. Test the system by speaking voice commands to Google Assistant and verifying that the home appliances respond accordingly. Check the Adafruit IO dashboard to confirm that the feeds are being updated correctly.

6.4 Experimental Results and Discussion

This section presents the results that validate the effectiveness of the prototype used in developing the IHAS. The prototype has demonstrated its ability to control various devices, including the motor, LED lights, bulb, and electric socket. The motor and LED lights are powered by a DC supply, while the electric bulb and the socket use an AC supply. Our work is further illustrated through Fig. 6.8a and b, showcasing images of a functional prototype. Figure 6.8a provides a front view of the prototype, highlighting the locations of the DC motor, LED, electric bulb, and electric socket. Similarly, Fig. 6.8b provides a top view of the prototype, providing clear identification of the various devices' positions.

By connecting to the Internet, users may keep an eye on and manage the prototype at any time and from any location. One of the highlights of the current work is that it can enable the person with disabilities to access the home equipment using voice commands. The artificial intelligence tools assist in recognizing the voice commands and IoT enables the controlling of appliances. Moreover, by creating a dashboard, users may operate and monitor their homes from anywhere at any time using a device that can connect to the Internet.

The functionality of the system is confirmed by testing on the IHAS prototype. The system manages and monitors all electrical appliances using Google Assistant, IFTTT, and Adafruit IO. In Fig. 6.9, the circuit connection using in the IHAS prototype is depicted. Raspberry Pi receives power supply from the USB-based power input. The supply drives the GPIO pins which enable or disable the relays available on the four-channel relay board. The relays are further connected to the different AC and DC loads as shown in Fig. 6.9.

Raspberry Pi is used to implement a Python program which reads the switch button conditions from Adafruit IO. Based on the condition of Adafruit IO, Raspberry Pi further controls the different home appliances through a four-channel relay board. Raspberry Pi connects with a pre-installed Wi-Fi network to communicate with the

(a) Front view (b) Top view

Fig. 6.8 Prototype of IHAS

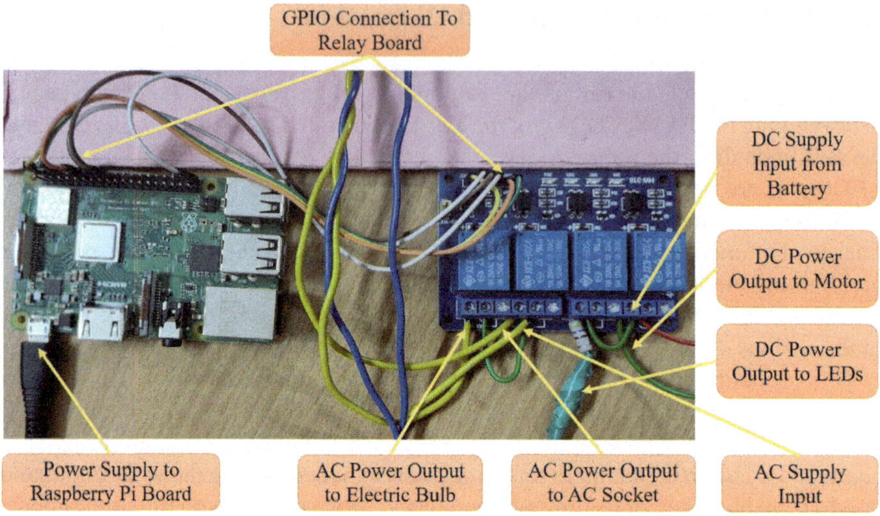

Fig. 6.9 Circuit connection for IHAS prototype

Adafruit IO server. Users can access Adafruit IO server, IFTTT server, and Google Assistant through their mobiles or personal computers. They need to create a user account and provide corresponding credentials to access the Adafruit IO and IFTTT services. A dashboard created in Adafruit IO provides an interface to manually control the different home appliances. Users may utilize the buttons on the GUI to turn on and off the appliances. Another way to control the home appliances is through voice commands. Google Assistant understands the predefined voice commands and interacts with IFTTT which further communicates with Adafruit IO. The value of input/output in Adafruit IO is continuously monitored by Raspberry Pi which, in turn, generates the commands to control relays connected with home appliances.

In previous studies, the existing home automation systems employ Arduino Uno, Arduino Mega, Raspberry Pi, or NodeMCU. We selected Raspberry Pi because of its advantages such as built-in Wi-Fi module and Bluetooth. On the other hand, other micro-controller-based platforms like Arduino require a separate Wi-Fi module to interface. The availability of built-in Wi-Fi on Raspberry Pi provides seamless connectivity with the Internet. Moreover, Raspberry Pi has more memory and data processing capability which enable it to integrate with multiple devices. We tested the IHAS prototype, by providing different voice commands to switch on and off the AC and DC loads. The sample list of the voice commands which can be used in the IHAS prototype is shown in Table 6.1. The list shows commands to activate and deactivate the devices in each of `English` and `Hindi` languages. We defined four commands for a single task. Out of four commands, two are in `English` and two are in `Hindi`. For eight tasks, we have total thirty-two commands as shown in Table 6.1. Figure 6.10 depicts the interface of Google Assistant while using the different commands for controlling some appliances. Figure 6.10a and b indicates

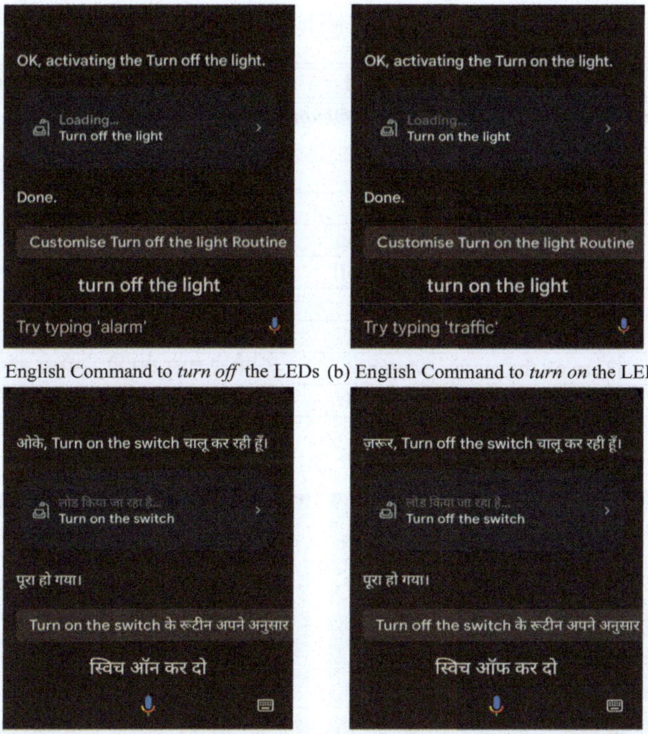

(a) English Command to *turn off* the LEDs (b) English Command to *turn on* the LEDs

(c) Hindi Command to *turn on* the socket (d) Hindi Command to *turn off* the socket

Fig. 6.10 Display of the commands while using Google Assistant

the display to switch on and off the LED lights, respectively, using Google Assistant. The user gives voice commands in `English` language. Similarly, the user can also speak in `Hindi` language and can activate an electric socket as shown in Fig. 6.10c and d.

We design a small dashboard on Adafruit IO to control and observe the switch conditions. Once the user logs in to Adafruit IO by providing account credentials, the dashboard can be used. We created four `Button` widgets corresponding to DC motor, LED lights, AC bulb, and AC socket (used to connect any home appliance) used in the IHAS prototype. We can change the button condition by pressing it. We have also connected Adafruit IO to Google Assistant through IFTTT service. Figure 6.11 depicts the dashboard consists of four `Button` widgets for controlling the electric loads of IHAS prototype. Figure 6.11a shows the Button's conditions when all the electric loads are active. On the other hand, Fig. 6.11b indicates only electric socket is active; and rest of the loads are in *Turn off* condition.

IFTTT triggers Adafruit IO in response to the voice command from Google Assistant. As a result, the condition of the corresponding `Button` widget on the dashboard

Table 6.1 Set of voice commands used in IHAS prototype

	English Command		**Hindi Command**	
Load Condition	*Voice command 1*	*Voice command 2*	*Voice command 3*	*Voice command 4*
Bulb On	Turn on the bulb	Bulb on	बल्ब चालू कर दो	बल्ब ऑन कर दो
Bulb Off	Turn off the bulb	Bulb off	बल्ब बंद कर दो	बल्ब ऑफ कर दो
Socket On	Turn on the switch	Switch on	स्विच चालू कर दो	स्विच ऑन कर दो
Socket Off	Turn off the switch	Switch off	स्विच बंद कर दो	स्विच ऑफ कर दो
LEDs On	Turn on the light	Light on	लाइट चालू कर दो	लाइट ऑन कर दो
LEDs Off	Turn off the light	Light off	लाइट बंद कर दो	लाइट ऑफ कर दो
Motor On	Turn on the motor	Motor on	मोटर चालू कर दो	मोटर ऑन कर दो
Motor Off	Turn off the motor	Motor off	मोटर बंद कर दो	मोटर ऑफ कर दो

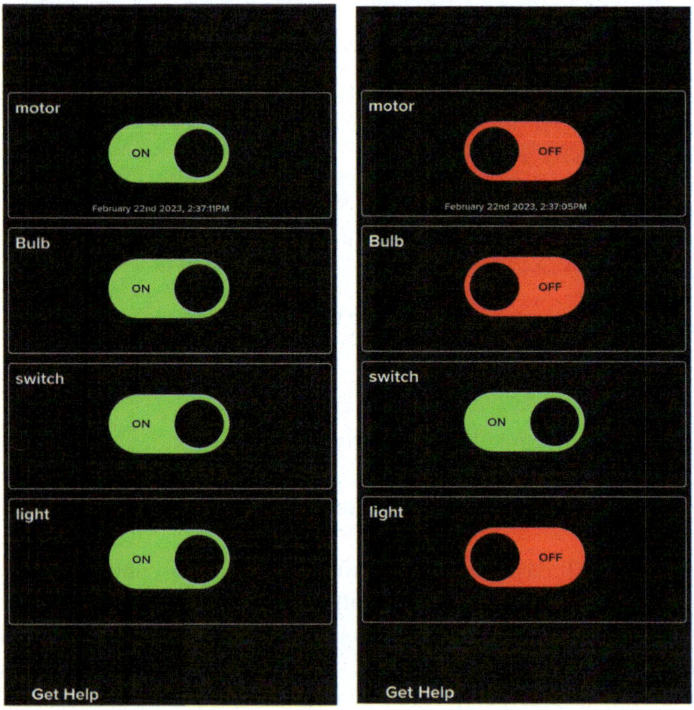

(a) Dashboard indicating all devices are in *turn on* condition

(b) Dashboard indicating only AC electric socket is in *turn on* condition

Fig. 6.11 Adafruit IO dashboard to control the IHAS appliances

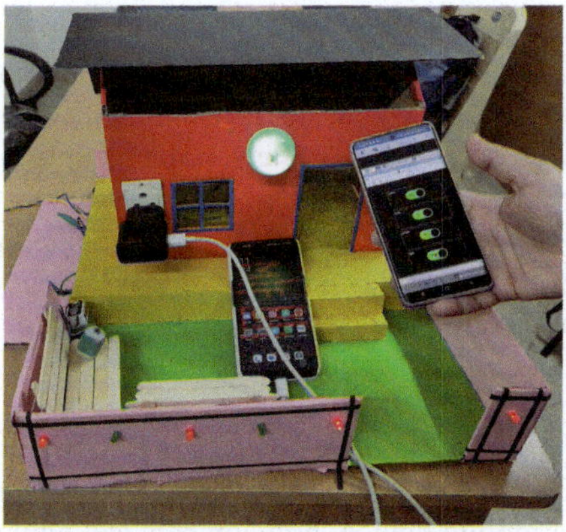

(a) IHAS prototype indicating all devices are in *turn on* condition

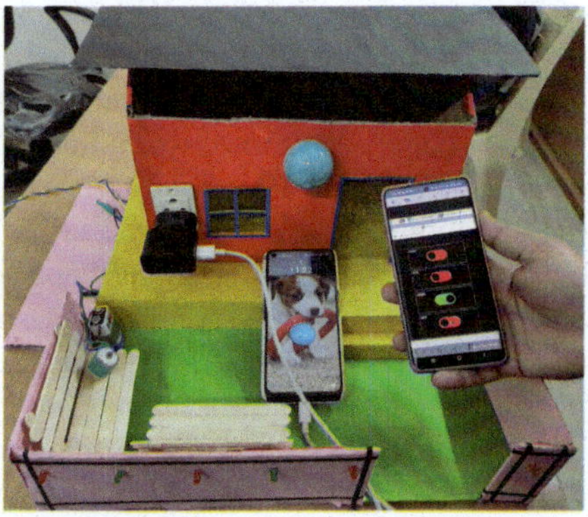

(b) IHAS prototype indicating only AC electric socket is in *turn on* condition

Fig. 6.12 IHAS prototype with active or inactive loads

changes. A Python program written in Raspberry Pi, continuously observe Adafruit IO. It detects the change in value of any button in Adafruit IO and consequently activates or deactivates the corresponding electric load. Figure 6.12 shows the IHAS prototype when different loads are in active or inactive conditions. The DC motor, LED lights, AC bulb, and AC socket in *Turn on* condition is shown in Fig. 6.12a. We connected a mobile phone in electric socket to charge it. Similarly, in Fig. 6.12b, it can be observed that only an electric socket is turned on and rest of the loads are in *Turn off* condition. While testing the prototype, we observed that sometimes it takes longer to activate or deactivate the device in response to a voice command. The reason for that is due to slow Internet connectivity, Raspberry Pi takes a little time to detect the change in Adafruit IO. In the presence of a good Internet connectivity, the developed system works efficiently.

The relay modules, in conjunction with the Raspberry Pi board and the IFTTT platform, enable seamless automation and management of various devices within the home. To control home appliances, the number of devices that can be independently controlled varies based on the channel capacity of the relay module. With a 4-channel relay module, up to four devices can be controlled individually. An 8-channel relay module allows for the independent control of up to eight devices, while a 16-channel relay module provides the ability to control up to sixteen devices separately.

It's worth noting that the actual number of devices controlled by IFTTT may also be influenced by the capabilities and limitations of the Raspberry Pi board being used. Factors such as the available GPIO pins and system resources should be taken into consideration when connecting and controlling multi-channel relay modules. In summary, by utilizing IFTTT with a Raspberry Pi board and a relay module, users can control multiple home appliances simultaneously, with the number of devices depending on the specific relay module's channel capacity.

The cost analysis of an IHAS includes purchasing and installing components like sensors, controllers, smart devices, and infrastructure. In our implementation, the Raspberry Pi board is the most expensive component, comprising about two-thirds of the total expenses excluding software costs. Despite the significant upfront investment, IHAS offers long-term benefits and cost savings. It optimizes energy consumption, leading to reduced utility bills over time. By intelligently managing heating, cooling, lighting, and other energy usage, the system enhances efficiency and generates savings. In conclusion, although IHAS installation involves higher initial costs, the potential for long-term energy efficiency and lower utility expenses justifies the investment for homeowners.

The IHAS accessible through mobile apps offers the convenience of remotely controlling and monitoring various aspects of a home's energy usage, thereby improving energy efficiency. With just a few taps on a mobile device, homeowners can adjust thermostat settings, turn off lights, and even power down appliances from anywhere in the world. This level of control enables individuals to actively manage energy consumption, ensuring that lights and devices are not left on unnecessarily. Additionally, IHAS can provide real-time energy usage data, allowing homeowners to identify areas of high energy consumption and make informed decisions to optimize efficiency. By harnessing the power of mobile apps for smart home automation, individuals have the tools at their fingertips to reduce energy waste, conserve resources, and contribute to a greener and more sustainable future.

6.5 Conclusion

The proposed home automation system offers homeowners a hassle-free and efficient living experience by utilizing IFTTT, Google Assistant, and Adafruit on a Raspberry Pi board. The system automates different home functions like lighting, security, and entertainment, and can be conveniently controlled using voice commands through Google Assistant. A mobile app is also available to control the system's devices and functions from a remote location. The Raspberry Pi board acts as a central hub that connects and manages various smart devices and sensors by using IFTTT, Google Assistant, and Adafruit IO.

The system is designed to address the daily issues people face, with a focus on affordability and convenience. Our goal was to design a user-friendly home automation system that can be conveniently operated from a distance. This system promotes energy efficiency by coordinating the operation of various household devices. With the conventional wall switches scattered throughout homes, managing appliances can be difficult for individuals with cognitive or mobility impairments. This proposed work offers a solution to these issues through the use of smartphones and remote-controlled home automation systems.

References

Abdulraheem AS, Salih AA, Abdulla AI, Sadeeq M, Salim N, Abdullah H, Khalifa FM, Saeed RA (2020) Home automation system based on IoT. Technol Rep Kansai Univ 62(5):2453–64

Akinbi A, Berry T (2020) Forensic investigation of google assistant. SN Comput Sci 1(5):272

Ali MH (2022) Smart home security: security and vulnerabilities. https://www.wevolver.com/article/smart-home-security-security-and-vulnerabilities

Asadullah M, Raza A (2016) An overview of home automation systems. In: 2016 2nd International conference on robotics and artificial intelligence (ICRAI). IEEE, pp 27–31

Baviskar J, Mulla A, Upadhye M, Desai J, Bhovad A (2015) Performance analysis of ZigBee based real time home automation system. In: 2015 International conference on communication, information & computing technology (ICCICT). IEEE, pp 1–6

Chaurasia T, Jain PK (2019) Enhanced smart home automation system based on internet of things. In: 2019 Third international conference on I-SMAC (IoT in Social. Mobile, Analytics and Cloud) (I-SMAC). IEEE, pp 709–713

Dash S, Choudekar P (2021) Home automation using smart devices and IoT. In: 2021 9th International conference on reliability, infocom technologies and optimization (Trends and Future Directions) (ICRITO). IEEE, pp 1–5

ElShafee A, Hamed KA (2012) Design and implementation of a WIFI based home automation system. Int J Comput Inf Eng 6(8):1074–1080

Gill K, Yang SH, Yao F, Lu X (2009) A zigbee-based home automation system. IEEE Trans Consum Electron 55(2):422–430

Gu Q (2006) RF system design of transceivers for wireless communications. Springer Science & Business Media

Gunge VS, Yalagi PS (2016) Smart home automation: a literature review. Int J Comput Appl 975(8887–8891)

Gupta MP (2018) Google assistant controlled home automation. Int Res J Eng Technol 5(5)

Hadwan HH, Reddy Y (2016) Smart home control by using raspberry pi and arduino uno. Int J Adv Res Comput Commun Eng 5(4):283–288

Hettiarachchi MC, Abeysiriwardhana WD, Aththanayaka AE, Panangala LD, Jayawardena C, Udara I (2022) A secure and intelligent smart home controlling system. In: 2022 3rd International informatics and software engineering conference (IISEC). IEEE, pp 1–6

Hoque MA, Davidson C (2019) Design and implementation of an IoT-based smart home security system. Int J Netw Distrib Comput 7(2):85–92

Jabbar WA, Alsibai MH, Amran NSS, Mahayadin SK (2018) Design and implementation of IoT-based automation system for smart home. In: 2018 International symposium on networks, computers and communications (ISNCC). IEEE, pp 1–6

Jiang L, Liu DY, Yang B (2004) Smart home research. In: Proceedings of 2004 international conference on machine learning and cybernetics (IEEE Cat. No. 04EX826), vol 2. IEEE, pp 659–663

Jie Y, Pei JY, Jun L, Yun G, Wei X (2013) Smart home system based on iot technologies. In: 2013 International conference on computational and information sciences. IEEE, pp 1789–1791

Jivani MN (2014) GSM based home automation system using app-inventor for android mobile phone. Int J Adv Res Electric Electron Instrum Eng 3(9)

John A, Santhosam IB (2014) Home energy management system based on ZigBee. Int J Inventive Eng Sci (IJIES) ISSN, pp 2319–9598

Khan F, ur Rehman A, Arif M, Aftab M, Jadoon BK, (2016) A survey of communication technologies for smart grid connectivity. In: 2016 International conference on computing, electronic and electrical engineering (ICE Cube). IEEE, pp 256–261

Kim KC, Wang DH, Han SB (2017) Home security system based on iot. J Korea Inst Electron Commun Sci 12(1):147–154

Kodali RK, Jain V, Bose S, Boppana L (2016) IoT based smart security and home automation system. In: 2016 international conference on computing, communication and automation (ICCCA). IEEE, pp 1286–1289

Krishna I, Lavanya K (2017) Intelligent home automation system using BitVoicer. In: 2017 11th International conference on intelligent systems and control (ISCO). IEEE, pp 14–20

Kung YF, Liou SW, Qiu GZ, Zu BC, Wang ZH, Jong GJ (2018) Home monitoring system based internet of things. In: 2018 IEEE International conference on applied system invention (ICASI). IEEE, pp 325–327

Li S, Xu LD, Zhao S (2015) The internet of things: a survey. Inf Syst Front 17:243–259

Luan H, Leng J (2016) Design of energy monitoring system based on IOT. In: 2016 Chinese control and decision conference (CCDC). IEEE, pp 6785–6788

Mittal Y, Toshniwal P, Sharma S, Singhal D, Gupta R, Mittal VK (2015) A voice-controlled multi-functional smart home automation system. In: 2015 Annual IEEE India conference (INDICON). IEEE, pp 1–6

Moon SJ, Shin H (2022) Agent for home server management in intelligent smart home network. Int J Internet Broadcasting Commun 14(2):225–230

Morshed NM, Muid-Ur-Rahman G, Karim MR, Zaman HU (2015) Microcontroller based home automation system using Bluetooth, GSM, Wi-Fi and DTMF. In: 2015 International conference on advances in electrical engineering (ICAEE). IEEE, pp 101–104

Munirathinam S (2020) Industry 4.0: industrial internet of things (IIOT). In: Advances in computers, vol 117. Elsevier, pp 129–164

Ovadia S (2014) Automate the internet with "if this then that" (IFTTT). Behav Soc Sci Libr 33(4):208–211

Pi R (2015) Raspberry pi 3 model b. https://www.raspberrypi.org

Reinisch C, Kofler MJ, Kastner W (2010) Thinkhome: a smart home as digital ecosystem

Richardson M, Wallace S (2012) Getting started with Raspberry PI. O'Reilly Media, Inc

Rubell B (2018) Welcome to Adafruit IO. https://learn.adafruit.com/welcome-to-adafruit-io/what-is-adafruit-io

Shabber SM, Bansal M, Devi PM, Jain P (2021) ihas: an intelligent home automation based system for smart city. In: 2021 IEEE International symposium on smart electronic systems (iSES)(Formerly iNiS). IEEE, pp 48–52

Sharma K, Reddy S (2022) Survey on smart home automation systems: tools, technology, architecture and communication interfaces. In: 2022 International conference on computing, communication, and intelligent systems (ICCCIS). IEEE, pp 406–413

Skeledzija N, Cesic J, Koco E, Bachler V, Vucemilo HN, Džapo H (2014) Smart home automation system for energy efficient housing. In: 2014 37th International convention on information and communication technology, electronics and microelectronics (MIPRO). IEEE, pp 166–171

Sriskanthan N, Tan F, Karande A (2002) Bluetooth based home automation system. Microprocess Microsyst 26(6):281–289

Tamakloe E, Kommey B (2022) A smart GSM-based home electrical appliances remote control system. IPTEK J Technol Sci 33(1):1–12

Ur B, Pak Yong Ho M, Brawner S, Lee J, Mennicken S, Picard N, Schulze D, Littman ML (2016) Trigger-action programming in the wild: an analysis of 200,000 IFTTT recipes. In: Proceedings of the 2016 CHI conference on human factors in computing systems, pp 3227–3231

Vaidya VD, Vishwakarma P (2018) A comparative analysis on smart home system to control, monitor and secure home, based on technologies like GSM, IOT, Bluetooth and PIC microcontroller with ZigBee modulation. In: 2018 International conference on smart city and emerging technology (ICSCET). IEEE, pp 1–4

Venkatesh K, Rajkumar P, Hemaswathi S, Rajalingam B (2018) IoT based home automation using raspberry Pi. J Adv Res Dyn Control Syst 10(7):1721–1728

Wilson C, Hargreaves T, Hauxwell-Baldwin R (2017) Benefits and risks of smart home technologies. Energy Policy 103:72–83

Yadav D, Singh Y, Gupta H (2018) Controlling of relay using Raspberry Pi via internet for home automation. Int J Adv Res Eng Technol 9(1)

Yunardi RT, Firdaus AA, Abdulloh MN, Nahdliyah SDN, Putra KT (2022) Relay module IoT devices for remote controlling of home automation system. In: 2022 2nd International conference on electronic and electrical engineering and intelligent system (ICE3IS). IEEE, pp 350–354

Chapter 7
Scaled Conjugate Gradient Backpropagation-Based Fault Analysis System for Induction Motor

Chaitanya Nimbargi, Anand Mahajan, Piyush Khadke, and Chetan B. Khadse

Abstract A fault analysis system for induction motor based on scaled conjugate gradient backpropagation is proposed in this paper. An induction motor is simulated in MATLAB with the specifications. Faults are created in the simulated induction motor as a part of data generation process for the artificial neural network. The fault data is used for the supervised learning which is the input for the training. As mentioned SCGB neural network is used as a learning algorithm. The data is divided into training, testing and validation data. The training performance is presented in the paper. The faults considered are symmetrical as well as unsymmetrical faults. The detected results for line to ground, line to line, double line, double line to ground, triple line, triple line to ground faults are analysed and presented in this paper.

Keywords Artificial intelligence · Induction motor · Neural network · Pattern recognition · Scaled conjugate gradient backpropagation

7.1 Introduction

The proper operation of induction motor and its application requires frequent maintenance and accurate and immediate fault detection. The fault detection in stator of induction motor and elimination of short circuits is difficult. As a result, various faults may occur, subsequently and continue until all vital electrical equipment has been damaged or destroyed. The fault type is not known. It is not required for routine maintenance, but it can be quite useful in the event of a fault. It will help in restoring

C. Nimbargi · C. B. Khadse (✉)
Dr. Vishwanath Karad MIT World Peace University, Pune, Maharashtra, India
e-mail: chetan.khadse@mitwpu.edu.in

A. Mahajan
Delft University of Technology, Delft, The Netherlands

P. Khadke
Black & Veatch, Pune, India

C. B. Khadse et al. (eds.), *Intelligent Methods in Electrical Power Systems*, Engineering Optimization: Methods and Applications, https://doi.org/10.1007/978-981-97-5718-3_7

the equipment to its original operating conditions in less time and in turn helping in cost saving for repairs. Discovery and removal of faults in an induction motor consists of flaws. As a result, fault detection and classification has been a constant challenge. Researchers, designers and maintenance engineers are all interested in this topic. Continuous fault free operation of Induction motor has become like a commodity in today's market-driven business environment. To keep the certain processes working, reliability of induction motor must be enhanced. It is critical that motor faults be recognized precisely, reliably and quickly. This can help the engineers restore the motor to working conditions in shorter span of time. Also help to keep the damage caused by the faults at minimum level.

The Hindawi published a paper, namely, "A Fault Analysis Method for Three-Phase Induction Motors Based on Spiking Neural P Systems". The paper showed the possibility of using Modified Fuzzy Reasoning Spiking Neural P Systems with Real Numbers to do fault Analysis of Three-Phase Induction Motors (Huang et al. 2021). In the application scenario of their proposed method, they have interfaced three-phase induction motor with a data acquisition followed by status monitoring which is fed to three different algorithms. Algorithm 1 performs pulse value reasoning, then algorithm to discover faulty symptoms and outputs probability with the help of forward prediction reasoning, whereas in algorithm 3 backward abductive diagnostic reasoning is performed. A case study, namely, "Fault Detection and Diagnosis in Induction Machines" (Marques et al. 2013) is published by Springer. In this case study Concerning to maintenance of induction machines, preventive maintenance is widely used currently. It consists in periodic inspections with the objective of replacing parts that are supposed to break after a certain number of hours. However, even with periodic inspections this type of maintenance presents disadvantages such as unneeded maintenance costs and in some situations catastrophic failures still likely to occur. Thus, the concept of condition-based maintenance (CBM) has emerged as an alternative to preventive maintenance. This type of maintenance is defined as the process of monitoring characteristics or parameters of a machine, in order to verify early changes and trends that can be used to indicate a fault situation or the need for maintenance. To determine if use of particular algorithm resulted in better performance, papers were reviewed on the use of different artificial neural network algorithms and its analysis. The literature indicates that scaled conjugate gradient algorithm can greatly contribute to technical feasibility of fault detection model with regards to pattern recognition.

The artificial neural network has proven to be an accurate method for the fault detection as per mentioned in the literature. The author of this (Lee et al. 2022) proposes a fault-detection system for induction motors considering the bearing faults, interturn shorts and broken rotor bars. The author has used artificial neural network (ANN) along with multiresolution analysis (MRA), correlation and fitness values-based feature selection (CFFS).

In Lee et al. (2019), author proposed a method based on convolutional neural network (CNN). The input data to the CNN is vibration signal data. This data is obtained from the induction motor experimental environment and then fault diagnosis is done by the CNN. The author of a paper (Sameh et al. 2020) has proposed the stator current analysis as input features-based diagnosis of induction machine by using the artificial neural network. In Mohamed et al. (2020) a method is proposed for detection of different Inter Turn Short Circuit (ITSC) faults using Artificial Neural Network (ANN) in an induction motor under consideration of various loading conditions. In one of the study from the literature (Navasari et al. 2018), discrete wavelet transform is used to detect the bearing damage by transient current analysis. The comparison of the signal sub-band frequency at normal bearings and during fault is done to determine the occurrence of damage, processing of transient current signals. Furthermore, artificial neural networks are used to provide information on classification of types of fault.

Along with the application in Induction motor, ANN with backpropagation algorithm is also used in various fields. In Pandey et al. (2021), transmission line fault detection and classification is reported which used neural network with backpropagation algorithm. The smart grid protection with electromagnetic field signals are reported in Khadse et al. (2021). In which electromagnetic field signals are generated from the current signals and then given to the neural network as an input for the training. In Khadse et al. (2017), scaled conjugate gradient backpropagation algorithm is used for the estimation of electromagnetic compatibility. Power quality monitoring is also one area in which backpropagation algorithms are used majorly (Khadse et al. 2015; Khadse et al. 2017).

An attempt is made here to check the accuracy of algorithm for induction motor fault detection and classification. The paper is organized as follows: Second section explains the proposed methodology. Third and fourth sections will elaborate the procedure of the experimentation. Fifth section will explain the results and its validation. Sixth section will explain the conclusion.

7.2 Proposed Methodology

When induction motors are running, several factors may lead to stator faults. 30–40% faults of the induction motor are stator faults. If it cannot be detected, faults at an early-stage stator failure will lead to more serious faults. In this paper, a study is carried out on how different stator faults affect different parameters in the induction motor. The faults are generated by using a fault generator block in MATLAB. The faults that are generated are triple line to ground (LLLG), triple line (LLL), double line to ground (LLG), double line (LL), line to ground (LG). It provides a short circuit path in a particular interval between two or more lines. The objective of the study is to simulate an induction motor and create faults for data generation. Once the data is generated, train an AI (Artificial Intelligence) system with data set of faults. After that, design an AI system to detect and classify faults. The various components used for this study are 3-phase voltage source (415 V), asynchronous machine (415 V, 18 kW, 4-pole, squirrel cage induction motor), fault generator block, nprtool, ANN block, interval test, low pass filter, bus selector, digital logic gates block, scope, powergui, constant block, V–I measurement block (Khadse et al. 2016).

The simulation as mentioned earlier was done on MATLAB Simulink. A three-phase source is used as the supply for the asynchronous machine or the induction motor. The fault generator produces a short circuit between two or more lines. It is connected to the stator of the machine to produce the faults. The asynchronous machine is a 415 V, 18 kW, 4-pole, Squirrel Cage Induction Motor. The quantities that are measured are the electromagnetic torque, rotor speed, rotor current of all the phases and the stator current of phase A. The load torque given to the induction motor is of 7 N m.

The following blocks are used for the simulation study in MATLAB.

Three-Phase Source: The three-phase source produces a balanced 3 Phase Voltage which is given as supply for the induction motor. It has a source resistance and source inductance of 0.893 and 16.58 mH respectively.

Asynchronous Machine: The asynchronous machine or the induction motor is of 415 V, 18 kW, 4 pole squirrel cage induction motor. It has a moment of inertia of about 0.05 kg m^2.

Fault Generator Block: The Fault generator blocks. It can produce the LG, LL, LLG, LLL, LLLG faults. Any phase can be used for the faults. The fault generator block produces a fault in a particular interval during the running of the motor.

Powergui Block: The powergui block allows you to choose one of these methods to solve your circuit: Continuous, which uses a variable-step solver from Simulink. Discretization of the electrical system for a solution at fixed time steps. Continuous or discrete phasor solution.

Low-Pass Filter Block: The block implements an analog Nth-order Butterworth filter with unit DC gain and varying cutoff frequency.

Interval Test Block: The interval test block was used for signal processing which outputs true (1) if the input is between the values specified by the Lower limit and Upper limit parameters. The block outputs false (0) if the input is outside those values. The output of the block when the input is equal to the Lower limit or the Upper limit is determined by whether you select the Interval closed on left and Interval closed on right check boxes.

7.3 Experimentation and Data Generation

The model for fault and data generation was made in Simulink. The model used in Fig. 7.1 was used for the purpose of data generation. The faults were modelled using the fault generation block. It models all the line faults. These faults are as follows:

- R phase to Ground fault
- B phase to Ground fault
- Y phase to Ground fault

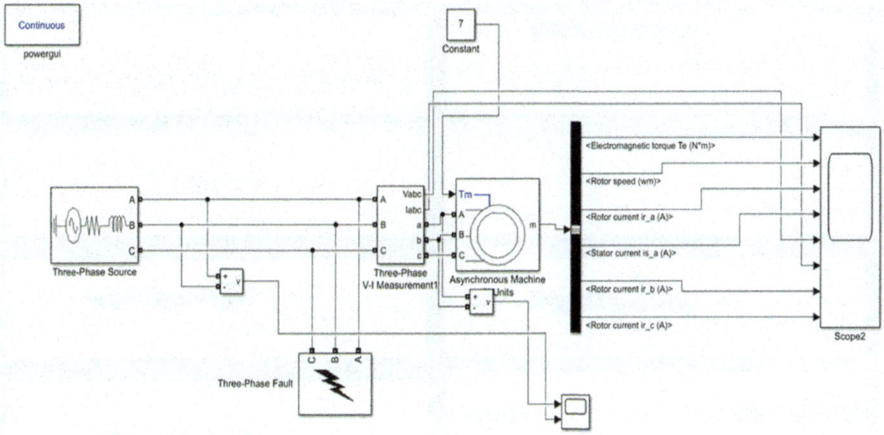

Fig. 7.1 Simulation diagram of the proposed methodology

- R phase to Y phase
- Y phase to B phase
- R phase to B phase
- R phase Y phase to Ground fault
- Y phase B phase to Ground fault
- R phase Y phase to Ground fault
- R phase Y phase B phase
- R phase Y phase B phase to Ground fault.

So, in total eleven faults have been considered and that have to be detected and classified. The classification is done such that when one of the 11 faults occurs the result will be 1 for that particular fault during the fault period and 0 when no fault period.

The waveforms have been observed and analysed. The waveforms analysed are the LG, LL, LLG, LLL, LLLG.

LG Fault

The waveforms for electromagnetic torque, rotor speed and rotor current are steady as shown in Fig. 7.2. The rotor speed before the fault occurs is 155.5. If we calculate the synchronous speed for the motor, we get it as 157.08 rad/s. If we calculate the slip, we get $s = 0.01$. If we calculate the rotor current frequency from the relation between rotor current frequency and stator frequency, we get the frequency of the rotor current as 0.5 Hz.

It is been observed that the electromagnetic torque has no oscillation where there is no fault as the torque is directly proportional to RMS value of rotor current. When the fault occurs the voltage for one line reduces to zero and the current going to the stator in that line decreases because the supply is the feeding the fault. Because of

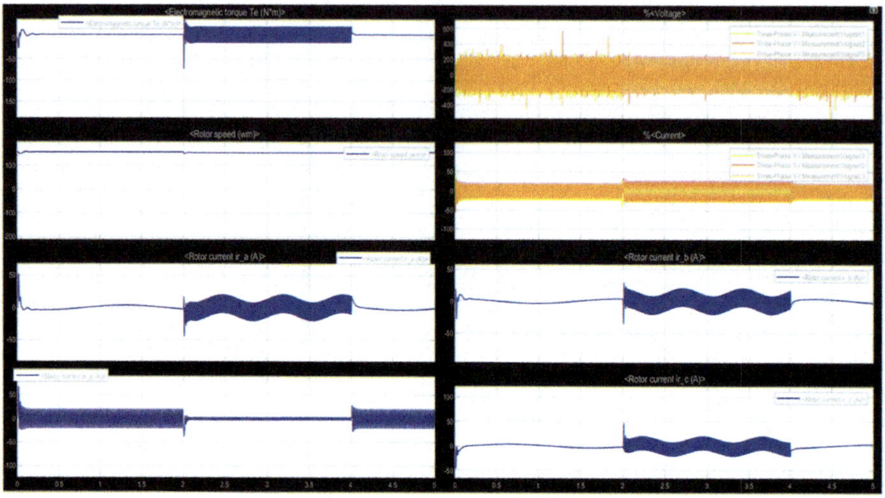

Fig. 7.2 Result for line to ground fault

the fault there is an unbalance between the voltages and the currents. This causes oscillations in the rotor current, rotor speed and the torque.

These oscillations can be explained by Fortescue's Theorem. Fortescue's Theorem states that a n-phase unbalanced system can be divided into n − 1 balanced systems and one zero sequence system. So, a 3-phase system can be divided into 2 balanced system and one zero sequence system. The positive sequence will produce RMF in the same direction as the balanced 3-phase system. The negative sequence current will produce RMF in the direction opposite to the balanced 3-phase system. Because of this reason, we can see from the rotor current that there are two sinusoidal rotor currents in opposite direction.

The sinusoidal current because of the positive sequence current will have a lower frequency because the RMF and the rotor are rotating in the same direction. The slip for the positive sequence RMF is approximately about 0.027. If we find the frequency of the rotor current, we get it equal almost to 1.5 Hz. For the negative sequence current the rotor is rotating in the opposite direction to the negative sequence RMF. Because of this reason the slip for the negative sequence RMF comes out to be almost equal to 2. So, the rotor current because for the negative sequence RMF is almost equal to 100 Hz. Both the positive sequence and negative sequence currents are opposing each other but do not cancel each other out because they are not of the same frequency. So, from the waveform of the rotor current we can see that there is an envelope which oscillates as a sine wave of lower frequency, and we have sine waves inside the envelope which consists of much higher frequency because of the negative sequence RMF. We can prove it by plotting the two currents as shown in the figure below (Fig. 7.3).

The torque consists of oscillations because the rotor current is a sine wave inside a sine wave. The frequency of the oscillating torque is also almost 100 Hz. The rotor

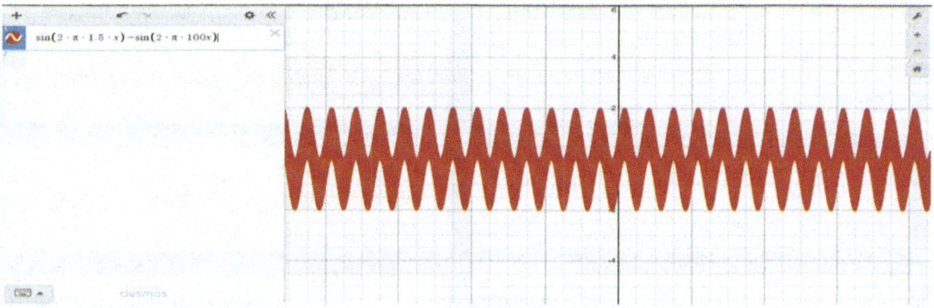

Fig. 7.3 Plot of current with higher frequencies

speed decreases to almost 153 rad/s because in the stable region, i.e. before the slip SM (slip for maximum torque) the torque is inversely proportional to speed. If we calculate the RMS value of the oscillating torque we see that the torque has increased from a value of 8–15.7 N m. So because of this reason the speed of the rotor decreases and the speed becomes oscillating.

LL Fault

For the LL Fault, we can see that the frequency of rotor current is increasing as we come towards the origin. This is because the rotor speed of the current is decreasing approximately linearly. After reaching zero speed the frequency starts increasing again. During no fault condition the torque is approximately 8 N m. But when the fault occurs the torque starts decreasing as the speed is decreasing because now the induction motor is in the unstable mode that is after SM. This SM will keep on changing because the frequency of the rotor current is changing which changes the rotor reactance X2. The torque is decreasing and it is less than the constant loading torque due to which the speed of the rotor starts decreasing.

Once the rotor speed reaches zero value the torque again starts increasing because the RMS value of rotor current increases. Now when the rotor is rotating in the opposite direction which means the torque due to friction and windage is supporting the electromagnetic torque. Due to this the rotor stops decelerating at the torque value of almost 6 N m. So now the rotor is rotating in the opposite direction. So after clearing the fault the slip will be greater than 1 and because of this the frequency will be greater than 50 (Fig. 7.4).

Once the rotor speed reaches zero value the torque again starts increasing because the RMS value of rotor current increases. Now when the rotor is rotating in the opposite direction which means the torque due to friction and windage is supporting the electromagnetic torque. Due to this the rotor stops decelerating at the torque value of almost 6 N m. So now the rotor is rotating in the opposite direction. So after clearing the fault the slip will be greater than 1 and because of this the frequency will be greater than 50.

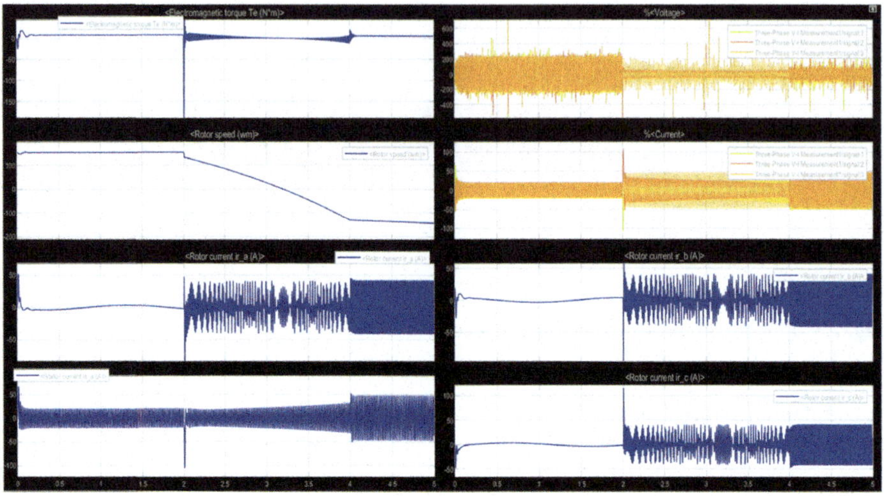

Fig. 7.4 Result for line to line fault

The value of the rotor current has increased because the value of the power factor has decreased because of the increase in frequency. Because of this reason, the value of the stator current has also increased.

LLG Fault

The LLG Fault is the same as the LL Fault only with the exception of stator voltage. The stator voltage in this case reduces to zero value during the faults (Fig. 7.5).

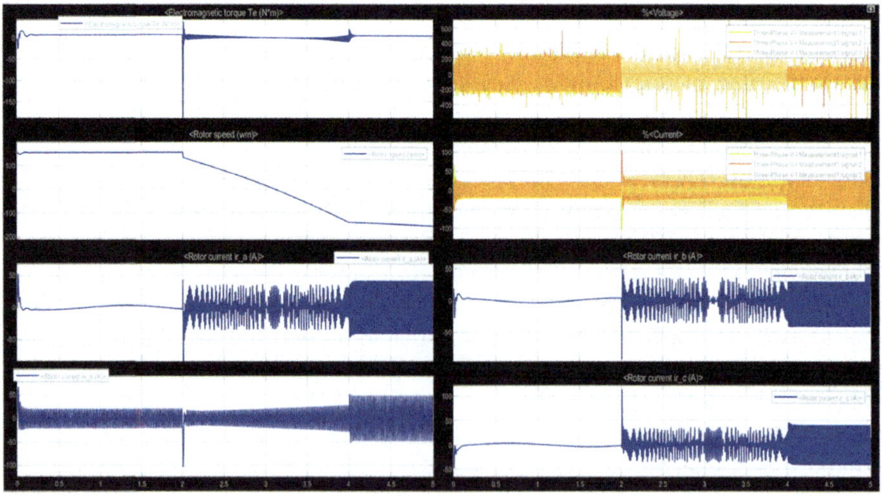

Fig. 7.5 Result for double line to ground fault

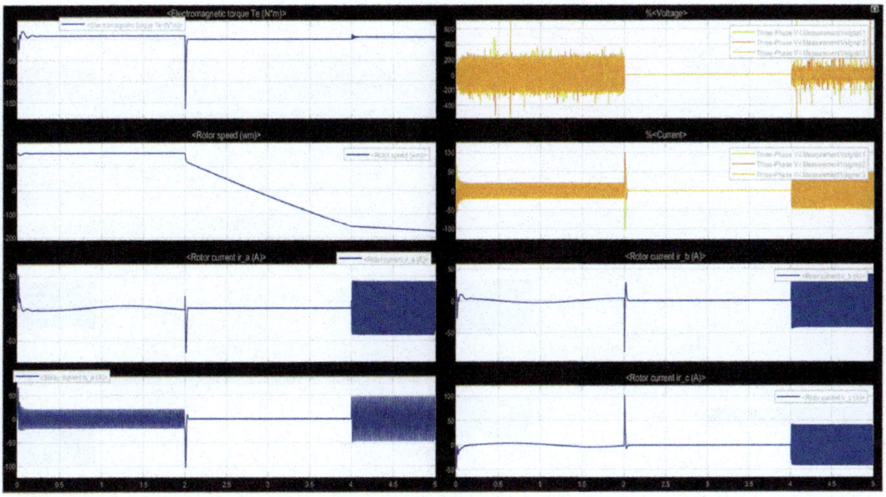

Fig. 7.6 Result for triple line fault

LLL Fault

In the LLL fault, all the lines are short circuited. Due to this the current going to the induction motor is reduced to zero as all of the current from the supply is feeding the fault. Because of zero current the torque produced is zero. Due to this, the rotor starts decelerating because of the constant load torque. After zero rotor speed the rotor starts rotating in the opposite direction. When the faults are cleared the rotor is rotating in the opposite direction and the torque is developed such that it balances the load torque. The torque developed after the faults is less than before the fault because in the case after fault clearance the friction and windage torque is supporting the electromagnetic torque. As the rotor is rotating in the opposite direction the slip is greater than 1 and because of that after fault clearance the frequency of the rotor current is more than 50 Hz (Fig. 7.6).

LLLG Fault

The waveforms of LLLG faults are the same as the LLL fault. As there is not much change related to it, it may be difficult to distinguish between these two faults (Fig. 7.7).

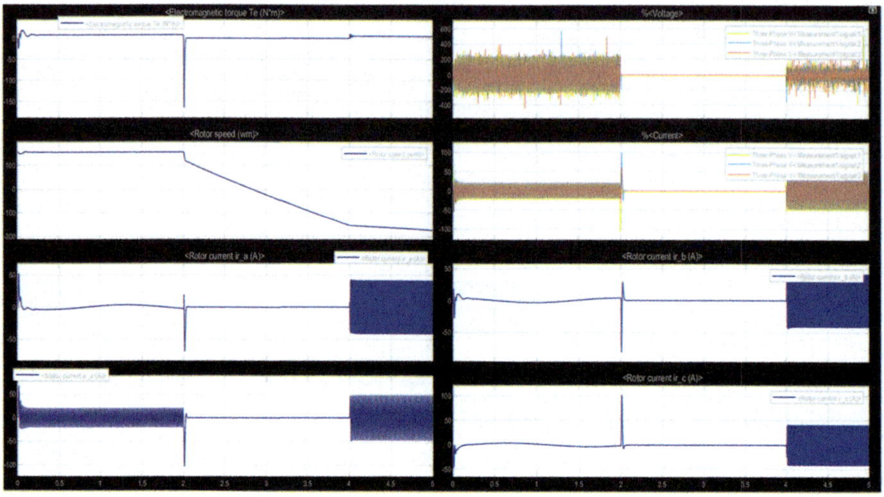

Fig. 7.7 Result for triple line to ground fault

7.4 The Neural Network Training

We have used the Scaled Conjugate Gradient Algorithm for training the Artificial Neural Network. 11 different ANN blocks were trained for classifying the faults. The samples for each fault were given to the ANN for training. The learning method used was the supervised learning method.

This means that we can give both the inputs and outputs for training purposes. The output is called target in this case. The output during faults is 1 and during no fault interval is 0. The voltages of the faults were given as samples to the ANN which was then trained according to the SCG algorithm.

The weights are adjusted in the steepest descent direction by the classic backpropagation method (negative of the gradient). This is the quickest-decreasing direction for the performance function. It turns out that, while the function reduces the most along the negative of the gradient, this does not always result in the quickest convergence (Khadse et al. 2016). A search is performed along conjugate directions in conjugate gradient algorithms, which generates generally faster convergence than steepest descent directions.

The backpropagation algorithm is similar to a proportional controller. In proportional controller if the gain is too high the oscillation increases and it diverges from the reference. If the gain is too low the time response is very low. Similarly, in backpropagation algorithm the variable step size acts as the proportional gain. If it becomes too high the oscillations increase. If the step size is too low the convergence is very slow. The gradient in the case of backpropagation algorithm is the derivative of the desired output and the system output. The difference between the desired and system output is the error. Because the network response to all training inputs must be computed numerous times for each search, this line search is computationally expensive. Moller [Moll93] devised the scaled conjugate gradient method (SCG) to circumvent the time-consuming line search (Figs. 7.8 and 7.9).

From the training results we can see that the best performance has been reached at the 14 epoch and after 6 validation checks the performance doesn't improve therefore, we stop at the 20th epoch. The gradient that we have reached is 3.77e^{-5} which is very close to the ideal gradient that is 1e^{-6}. The target value for performance should be 0 and our value is very close to it. It is 0.0311.

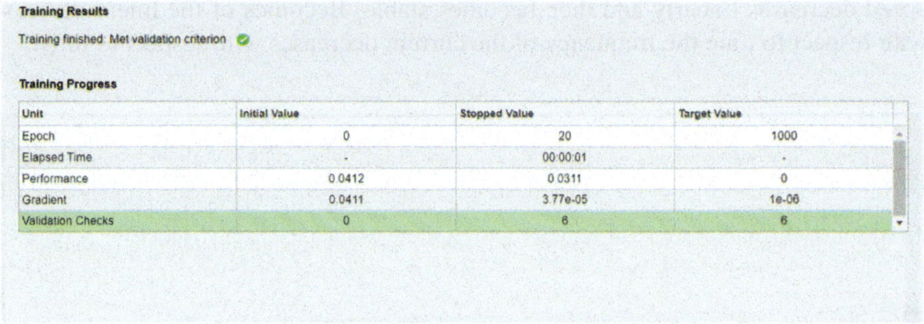

Training Results

Training finished: Met validation criterion

Training Progress

Unit	Initial Value	Stopped Value	Target Value
Epoch	0	20	1000
Elapsed Time	-	00:00:01	-
Performance	0.0412	0.0311	0
Gradient	0.0411	3.77e-05	1e-06
Validation Checks	0	6	6

Fig. 7.8 Training results and the parameters proof

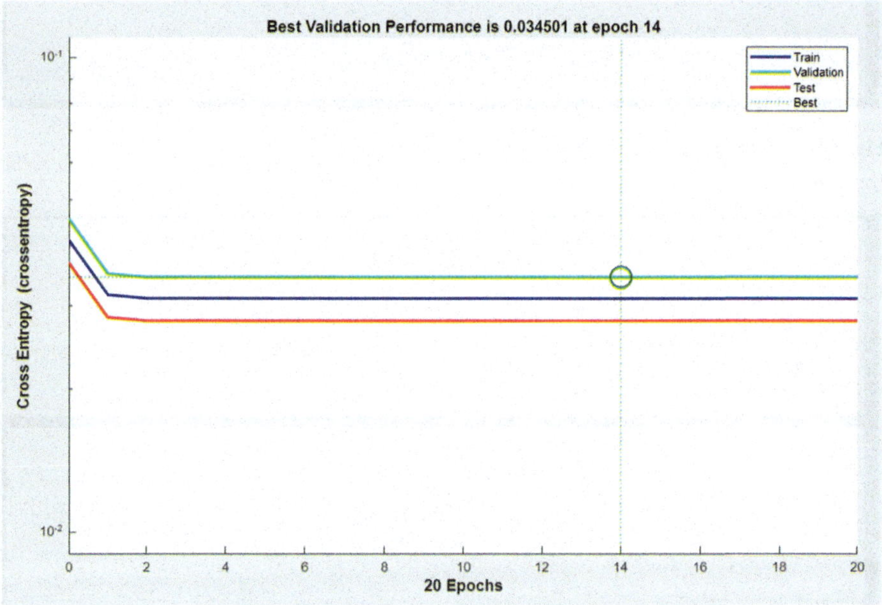

Fig. 7.9 Graph of the training iterations

7.5 The Results of the Proposed System

The results show the effect of the positive sequence and negative sequence on the waveforms. In most of the cases the speed of the rotor drops. In LG fault the rotor speed drops and recovers again. In all other faults the rotor speed drops but after clearing the fault the rotor starts rotating in the opposite direction. The torque decreases because now the friction and windage torque is now supporting the electromagnetic torque (Figs. 7.10, 7.11, 7.12 and 7.13).

The LL and LLG faults are similar the only difference is between the voltages. The speed decreases linearly and then becomes stable. Becomes of the linear decrease with respect to time the frequency of the current decreases with respect to time.

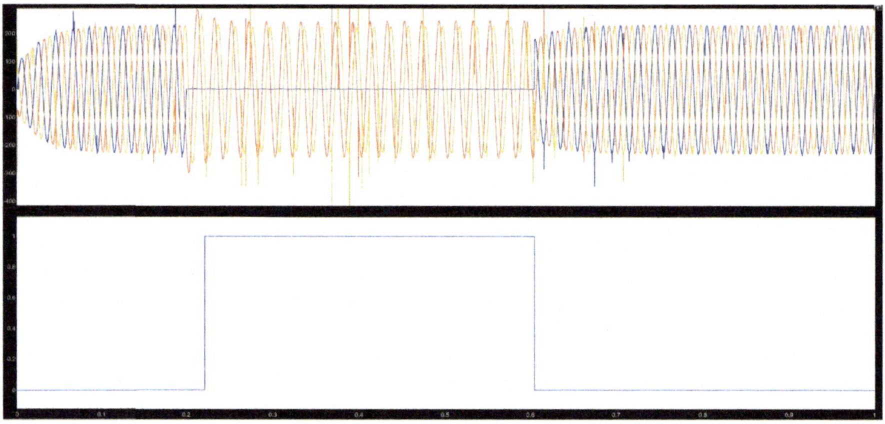

Fig. 7.10 Neural network result for R phase to ground fault

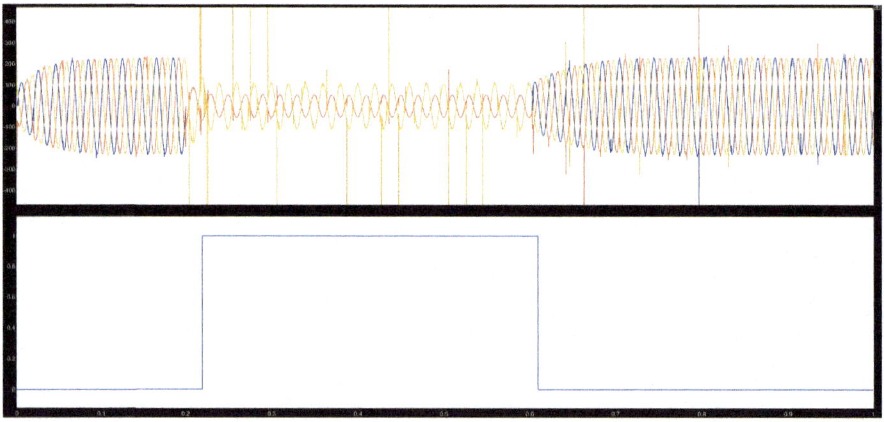

Fig. 7.11 Neural network result for R phase to Y phase fault

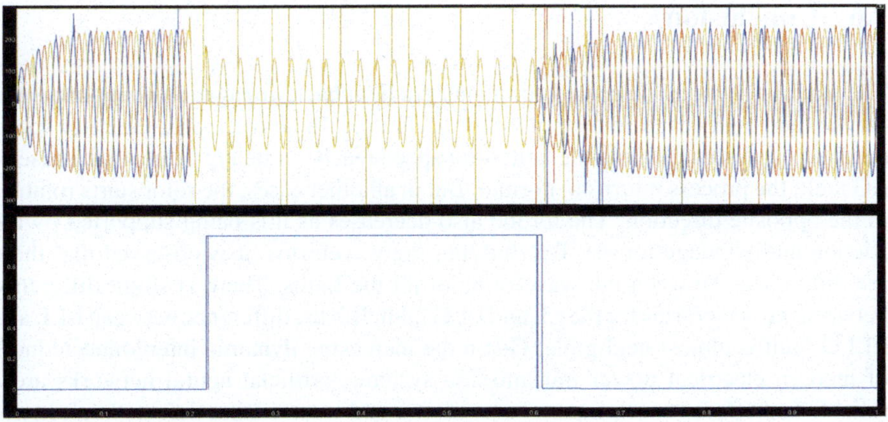

Fig. 7.12 Neural network result for R phase Y phase to ground fault

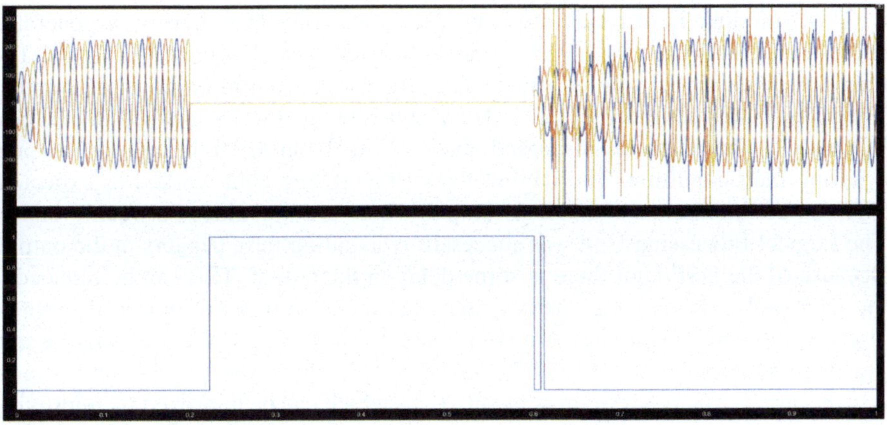

Fig. 7.13 Neural network result for R phase to Y phase to B phase fault

The LLL and LLLG waveform have the same waveforms. In the LLL and LLLG all the voltages and the currents reduce to zero. Because of this all the parameters i.e. torque, speed, rotor current all reduce to zero. Because of no current in the rotor there is no torque produced. Therefore, all the parameters are zero. Because of zero electromagnetic torque the rotor rotates in the opposite direction.

From the results, we can see that the classification and detection has been done successfully. For a particular fault the output is 1 during the fault and during no fault condition this value is 0. The classification has also been done successfully. The classification has been done such that out of 11 faults the scope shows which fault has occurred.

7.6 Conclusion

The neural network with scaled conjugate gradient backpropagation is implemented for the fault analysis of induction motor accurately. In all the faults, the rotor speed decreases and after clearing the fault rotor speed reaches a steady value. In the case of LG fault, the process returns to normal. But in all other cases, the rotor starts rotating in the opposite direction. The torque also decreases as it is being supported by the friction and windage torque. By studying the waveforms, it is observed that there are differences between the waveforms of all the faults. There is slight difference between the waveforms for the LL and LLG fault. But the difference between LLL and LLLG fault is almost negligible. Given the increasing dynamic interconnectedness of modern electrical power transmission systems, artificial neural networks are a reliable and effective technique of identifying and categorizing defects in electrical power transmission lines. Before using an artificial neural network in a practical application, the network's performance, as well as the topology and learning process of the network, should be extensively investigated.

This Simulink model was used for data generation. The data was generated successfully. This data was given to the pattern recognition tool in MATLAB. The data was given in the form of arrays. The target was given in the form of 0 and 1. The training was done successfully. The results obtained were obtained in the form of 0 and 1. The output as discussed above correctly shows 0 during no fault and 1 during fault condition. The Digital Signal Processing Unit worked as expected. The output shows clear 0 and 1 as output. There are no values between 0 and 1. The Logical Processing Unit was successfully avoiding false triggers in the output. Because of the DSP Unit there is some delay in the output. This can be reduced if the DSP unit is made faster. This system can be easily used in practical settings. None of the components in the model is ideal. They are all practical sources and practical induction motor. The faults have been classified and detected with almost 93% accuracy. There is a slight delay of 20 ms which can be improved by improving our data processing unit. The system can also be used for practical systems as none of the components is ideal.

References

Huang Z, Wang T, Liu W, Valencia-Cabrera L, Pérez-Jiménez MJ, Li P (2021) A fault analysis method for three-phase induction motors based on spiking neural P systems. Complexity 2021:19. Article ID 2087027. https://doi.org/10.1155/2021/2087027

Khadse CB, Chaudhari MA, Borghate VB (2016) Conjugate gradient back-propagation based artificial neural network for real time power quality assessment. Int J Electr Power Energy Syst 82:197–206

Khadse CB, Chaudhari MA, Borghate VB (2017) Electromagnetic compatibility estimator using scaled conjugate gradient backpropagation based artificial neural network. IEEE Trans Industr Inf 13(3):1036–1045

Khadse C, Patharkar AA, Chaudhari BS (2021) Electromagnetic field and artificial intelligence based fault detection and classification system for the transmission lines in smart grid. Energy Sources, Part A: Recover, Util, Environ Eff. https://doi.org/10.1080/15567036.2021.1948637

Khadse CB, Chaudhari MA, Borghate VB (2015) Comparison between three back-propagation algorithms for power quality monitoring. In: 2015 Annual IEEE India conference (INDICON), New Delhi, India, pp 1–5. https://doi.org/10.1109/INDICON.2015.7443766

Khadse CB, Chaudhari MA, Borghate VB (2016) A laboratory set-up for power quality disturbance generator and real time power quality monitoring. In: 2016 IEEE international WIE conference on electrical and computer engineering (WIECON-ECE), Pune, India, pp 61–64

Khadse CB, Chaudhari MA, Borghate VB (2017) Comparison of seven backpropagation algorithms for three phase power quality assessment. In: TENCON 2017—2017 IEEE region 10 conference, Penang, Malaysia, pp 2548–2553. https://doi.org/10.1109/TENCON.2017.8228291

Lee J-H, Jae-Hyung P, In-Soo L (2019) Fault diagnosis of induction motor using convolutional neural network. Appl Sci 9(15):2950. https://doi.org/10.3390/app9152950

Lee, C-Y, Meng-Syun W, Guang-Lin Z, Truong-An L (2022) Application of ANN in induction-motor fault-detection system established with MRA and CFFS. Mathematics 10(13):2250. https://doi.org/10.3390/math10132250

Marques M, Martins J, Pires VF, Jorge RD, Mendes LF (2013) Fault detection and diagnosis in induction machines: a case study. In: Camarinha-Matos LM, Tomic S, Graça P (eds) Technological innovation for the internet of things. DoCEIS 2013. IFIP advances in information and communication technology, vol 394. Springer, Berlin, Heidelberg. https://doi.org/10.1007/978-3-642-37291-9_30

Mohamed M, Mohamed E, Mohamed A-A, Abdel-Nasser M, Hassan MM (2020) Detection of inter turn short circuit faults in induction motor using artificial neural network. In: 2020 26th conference of open innovations association (FRUCT), Yaroslavl, Russia, pp 297–304. https://doi.org/10.23919/FRUCT48808.2020.9087535

Navasari E, Asfani DA, Negara MY (2018) Detection of induction motor bearing damage with starting current analysis using wavelet discrete transform and artificial neural network. In: 2018 10th international conference on information technology and electrical engineering (ICITEE), Bali, Indonesia, pp 316–319. https://doi.org/10.1109/ICITEED.2018.8534749

Pandey A, Gadekar PS, Khadse CB (2021) Artificial neural network based fault detection system for 11 kV transmission line. In: 2021 International conference on advances in electrical, computing, communication and sustainable technologies (ICAECT), Bhilai, India, pp 1–4

Sameh M, Tarek A, Yassine K (2020) Bearing and rotor faults detection and diagnosis of induction motors using statistical neural networks. In: 2020 20th international conference on sciences and techniques of automatic control and computer engineering (STA), Monastir, Tunisia, pp 77–81. https://doi.org/10.1109/STA50679.2020.9329334

Chapter 8
Ice Thickness Control Circuit to Automate the Milk Chilling System

Kalash Nikhar, Ruchir Nimgaonkar, Shivankar Patil, Abhidnya Kadu, and Amit Nehete

Abstract Milk collection is a common logistical challenge in dairy industry due to the remoteness of the villages and the road conditions. Milk chilling system plays a crucial part in the milk collection process from collection centers to the factories for processing and packaging. The total time required for milk to reach the chilling center after the farmer milks cow can be as high as 7–8 h. During this time, bacterial growth may spoil the milk leading preservation of milk to a critical issue without/insufficient cooling. The paper presents the development and implementation of an Ice Thickness Controller circuit that starts the cooling as milk can is placed in the chilling system and stops the cooling as soon as milk is cooled to desired temperature and ready for the transportation. An Ice Thickness Sensor is designed using the principle of conductivity. Outcome of the work is an automated milk cooling process that enables energy saving by running compressor on and off on reaching the thresholds of the temperatures ensuring the required ice level.

Keywords Conductivity sensor · Micro controller · Milk cooling

8.1 Introduction

The expansion of India's rural economy has been greatly aided by the dairy industry. India produces 23% of the world's milk, making it the greatest producer in the world, with many of its people relying on milk as a primary source of income. Indian dairies buy milk from local farmers at village collection centers, and then sell the milk or use it to make dairy products (Matheson 2015; Ministry of Fisheries, Animal Husbandry and Dairying 2022).

K. Nikhar · R. Nimgaonkar · S. Patil · A. Nehete (✉)
Dr. Vishwanath Karad MIT World Peace University, Pune, India
e-mail: amit.nehete@mitwpu.edu.in

A. Kadu
Eindhoven University of Technology, Eindhoven, Netherlands

C. B. Khadse et al. (eds.), *Intelligent Methods in Electrical Power Systems*, Engineering Optimization: Methods and Applications, https://doi.org/10.1007/978-981-97-5718-3_8

Milk collection is a common logistical challenge in dairy industry of most of the developing countries, due to the remoteness of the villages and the poor road conditions. The total time required for milk to reach the chilling center after the farmer milks the cow can be as high as 7–8 h. During this time, bacterial growth may spoil the milk leading preservation of milk to a critical issue without/insufficient cooling. While the fatty acid profile of the delayed chilled milk undergoes significant alterations, the lactose and pH of the delayed chilled raw milk are found lower than those of the instantly chilled milk. The peroxide value and antioxidant capacity of delayed chilled raw milk are noticed to be higher. It is recommended that milk collection system in the developing countries should be improved to avoid any delay in between milking and chilling of raw milk (Ajmal et al. 2018). The microbial quality of milk starts a farm level. Milk is sterile at secretion in the udder but is colonized by bacteria before it leaves the udder (Evans 2016; Paludetti et al. 2018). The temperature of milk expelled from the udder is approximately 35 °C; to prevent microbial growth, rapid cooling, and storage to 4 °C is necessary (Burke et al. 2018).

One of the main challenges in small holding dairy production systems is the commencement of cold-chain from the point of milk production. Prakash and Ravindra (2018) and Zakeri et al. (2023) underlines the need of energy efficiency in chilling of small volumes of milk. Modeling and testing of an ice bank for milk cooling after milking is presented in Luís et al. (2018). The energy balance equations and heat transfer arrangements utilized in the modeling provided good approximations for predicting the parameters linked to ice formation. However, more realistic parameters are required to be used for obtaining better accuracy of ice prediction. (Ghewade et al. 2018) underlines the necessity to study bulk milk cooling systems and improve them to reduce chilling time and energy consumption within desired limits. In order to create a dynamic model for bulk milk coolers, the authors used a vapor compression system with R22 as the refrigerant.

Nearly 400 million liters of milk are produced every day in India, of which 100 million liters are used for processing. Additionally, according to the National Dairy Development Board, only 20 million liters of the 100 million liters are refrigerated at the village level. Milk chilling system plays a crucial part in the milk collection process from collection centers to the factories for processing and packaging. Dairies avoid distant areas because it is expensive to install a milk chiller and there is no method for farmers to deliver their milk to larger dairies. They ultimately sell it at the neighborhood market, where there is a high risk of spoilage, or they sell it to a middleman who will take a commission. Therefore, these low-cost milk-chilling units are of prime importance in the context (Thaker and Tripathi 2007). Micro-can Chiller and Rapid Milk Chiller are the among the most popular milk cooling milk solutions used to cool the milk from its harvesting temperature of 35 °C to storage temperature of 4 °C (Buddhi and Ravi 2019). One of the important functions of this milk chilling system is to know the volume of ice formed inside. This information is used to program the charging and discharging circuit.

Despite of large existence of conventional milk cooling processes, development of any specific Ice thickness controller for the ice formed in the milk chilling systems is yet to be noted in the literature that leaves the gap for undertaking the research in this area.

8.2 Milk Chilling Process

The initial temperature of milk when collected in cans is usually about 35 °C (Paludetti et al. 2018; Burke et al. 2018). The chilling process begins as soon as milk cans are placed in the system. Liquid is pumped into the pipes, and then compressor starts in order to cool the same. It starts by pumping liquid into the pipes of the system and then compressor starts. The compressor cools the liquid inside the pipes. Hence, there is ice formation on the pipes, and thickness of the ice formed increases as the temperature starts falling down. The ice formed then cools the water in the system.

The cool water is then sprinkled on the cans to cool the milk inside the cans. Then as the ice formation is complete, the milk temperature reaches 4 °C. The milk chilling process is therefore complete and milk is ready to be transported. Figure 8.1 shows the existing system of the milk chilling process.

8.2.1 Problem Identification

In the process of ice formation, the existing system is used to regulate ice formation by measuring time taken to form the ice. The system was manually monitored and sometimes there would be more/less ice formation in system. One of the important functions of the milk chilling system is to know the volume of ice formed inside. The ice formation depends on multiple factors one of them being TDS (Total Dissolved Solids). TDS leads changes in freezing temperature of water. This affects the Ice Volume. Finding strategies to estimate the volume of ice created is necessary because it is difficult to quantify the actual volume. Also, automation of the compressor operation is essential to regulate the ice formation.

8.2.2 Proposed Solution

Develop an ice thickness controller circuit to automate the process such that it will automatically start and stop the cooling to enable the energy saving.

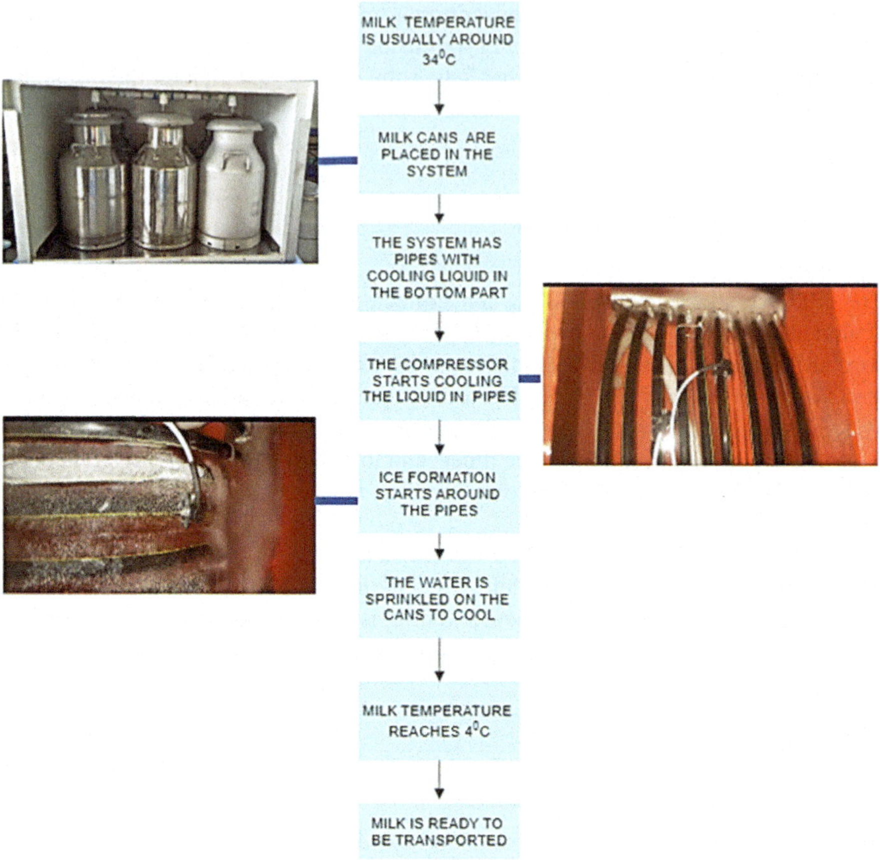

Fig. 8.1 Block diagram of milk chilling process

To automate the process, a sensor has been designed which will sense the ice thickness in the system. Then we have programmed the control circuit in such a way that it will stop cooling as soon as it reaches the required thickness level. Hence, our objective to automate the system so that milk remains chilled and is ready for transportation at collection centers is achieved.

The project has been done in two stages; first stage was to work on the ice thickness sensor, which includes designing the sensor and implementing it and the second stage was to work on the control circuit and mechanism of the Ice Thickness Controller.

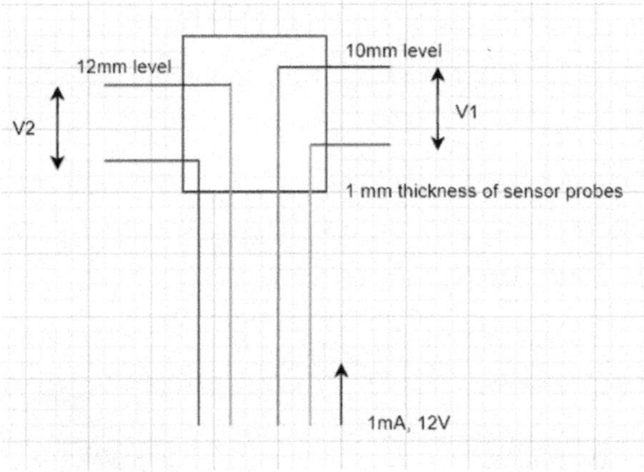

Fig. 8.2 Electrical design of a sensor

8.3 Controller Circuit Design

8.3.1 Sensor Design

This sensor works on the principle of conductivity, wherein the ice acts as an insulator once the desired level of ice is attained (Thirstrup and Deleebeeck 2020). The sensor is provided with four probes with two different levels (10 and 12 mm) as shown in Fig. 8.2. V1 indicates the voltage difference between 10 mm and V2 indicates voltage difference between 12 mm.

Sensor receives an input signal of 12 V, 1 mA. Material used for probes—SS304. Type 304 stainless steel is an austenitic stainless steel from the T 300 Series. Material used for coating on probes—Tin plating, tin plating is the process of depositing a coating of solderable tin plating onto the surface of a material via an electrical current. Material used for molding—Polypropylene (PP), polypropylene is easy to mold despite its semi-crystalline nature, and it flows very well because of its low melt viscosity (Bell 2021). Number of cores used for probe wires-4. We have decided Red and Yellow (10 mm) and Green and Black (12 mm), 10 and 12 mm are the level of ice thickness. Standard length of the wire for probe is 10 ft. Figure 8.3 and 8.4 shows the design and actual pictures of the developed sensor respectively.

8.3.2 Controller

Block Diagram

Figure 8.5 shows the block diagram of Ice Thickness Controller. The microcontroller is given a power supply of 9 V. Output signal from the ice thickness sensor

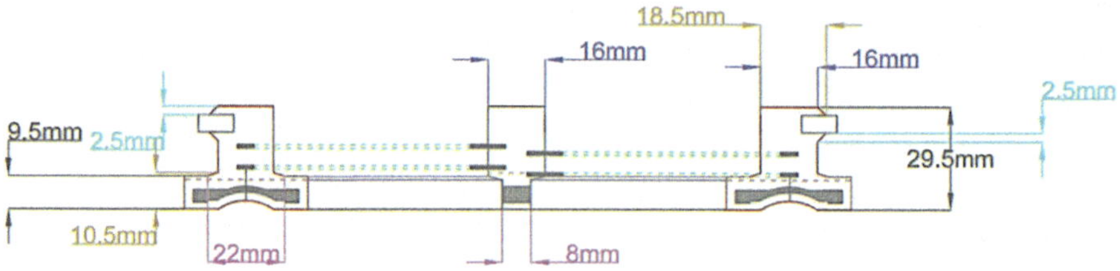

Fig. 8.3 Both side view of a sensor

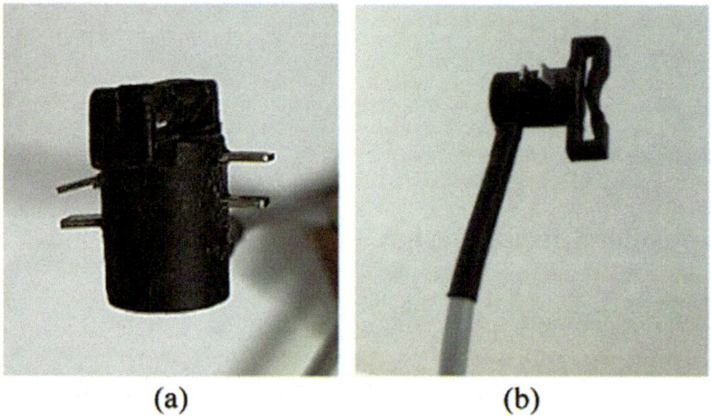

Fig. 8.4 Actual photos of a developed sensor

is transmitted to microcontroller. Microcontroller sends the signal to relay to turn the compressor On and Off as per the logic defined. LED light will show the status depending on the condition of the compressor.

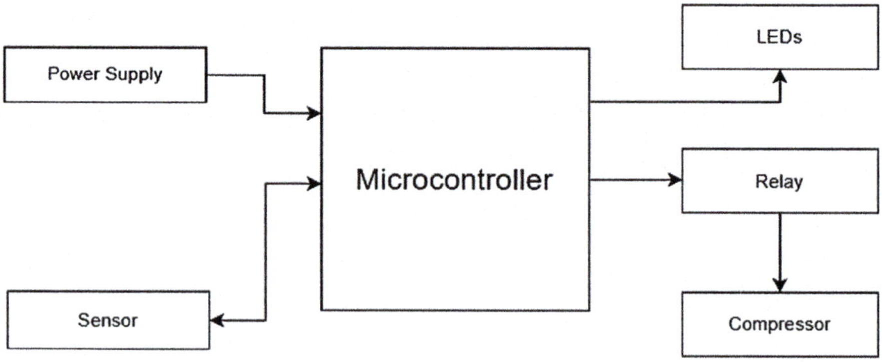

Fig. 8.5 Block diagram of ice thickness controller circuit

The power source consists of step-down transformer which is connected to full bridge rectifier that converts AC to DC. In this circuit, separate sub-circuits have been created for 10 mm probes and 12 mm probes. When 10 mm probes are being sensed 12 mm probes remain inactive, as the circuits are isolated. This configuration avoids errors in probe readings. LED configuration part of the circuit is responsible for indicating ice formation levels with the help of three LEDs. Red LED indicates zero ice formation. Orange LED indicates 10 mm ice formation. Green LED indicates 12 mm ice formation. Relay configuration part helps to switch on/off the compressor. Here, Darlington pair of transistors is used to get high current gain to supply it to the relay. Microcontroller used in the circuit is Atmega8. Microcontroller helps to control the output of the relay and for indication of ice formation levels via LEDs. Figure 8.6 illustrates the connection of various components of these blocks on circuit board.

Software. The C language code for the logical operation of Ice Thickness Controller (refer appendix) is created in Atmel Studio 6. Atmel Studio 6 is designed for hardware developers to develop and debug microcontroller applications. Applications built in C/C++ and assembly language run smoothly on this integrated development environment, which employs the Microsoft Visual Studio shell.

The code was dumped using the eXtreme Burner. A virtual system modeling and circuit simulation application is called Proteus. It replicates all of the supported processor's peripherals, including input/output ports, interrupts, timers, USARTs, and more. Circuit is developed in Proteus: A Virtual System Modeling and circuit simulation application.

8.4 Hardware Implementation

The Ice Thickness Sensor sends signal wirelessly from the sensor to the operator's computer and displayed in real-time as a cross-sectional image. The microcontroller enables the system designer to balance processing speed and power consumption. The transformer we have used is the step-down transformer which converts 230–9 V 60 mA and 14 V, 125 mA.

Voltage regulator IC 7805 adds a provision for a heat sink. The L7809 voltage regulator provides 9 V Positive voltage as output. Resistors help in controlling the flow of the current in the circuit. Capacitor is used for storing electrical energy, consisting of two conductors in close proximity and insulated from each other. The 1N4748 diode is ideal for use in power supplies and converter circuits. The three LEDs used were red, orange, and green which helps us in indicating the level of ice formation. Relays are a crucial component for the protection and switching of several control circuits and other electrical parts.

The list of hardware components used in Fig. 8.7 is presented in Bill of material (BoM) in Table 8.1.

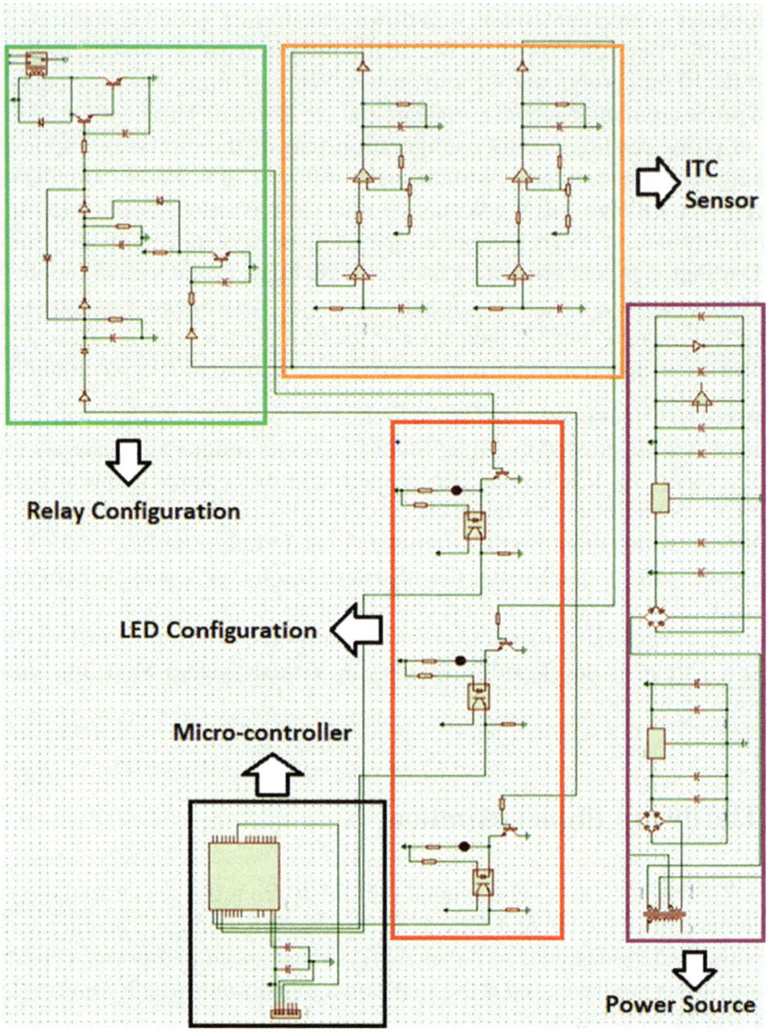

Fig. 8.6 Overall circuit of the ice thickness controller

8.5 Results and Discussion

8.5.1 Test Cases

See Table 8.2.

Table 8.1 Bill of material (BoM) used in the developed circuit

Component	Qty.	Specifications	Cost/unit (Rs.)	Price (Rs.)
Microcontroller (Atmega 8)	1	8-Bit microcontroller	48	48
LED	3	1.8–3.3 V	5	15
Transformer	1	230–9 and 14 V	250	250
Diode	4	4148	5	20
Capacitor	6	47 micro-F to 470 micro-F	10	60
Relay	1	10 A/120–12 V DC	100	100
Resistor	20	120 Ω–10 kΩ	3	60
Voltage regulator	2	LM7805, LM7809	100	200
Transistor	5	547	2	10
Ice thickness sensor	1	12 V/1.2 mA–5 V	1500	1500
Total cost (Rs.)				2263

Table 8.2 Test cases after the implementation

LED			Ice formation		Compressor status
Red	Orange	Green	10 mm	12 mm	
1	0	0	0	0	On
0	1	0	1	0	On
0	0	1	1	1	Off

8.5.2 *Performance at Different TDS*

In the milk chilling systems, sometimes different salts of sodium and magnesium are added in water to speed up the ice formation. These salts increase the TDS level of water. Therefore, to test the performance of sensor in different TDS conditions, several tests were carried out. The results obtained are presented below.

TDS Level 0–200 ppm

From the graph in Fig. 8.8, we can observe that when Ice formation reaches 10 mm, the V1 doesn't increase drastically, as the supply is cut off, however when 12 mm is reached V2 supply gets cut off and V2 value is maximum (Table 8.3).

TDS Level 400 ppm

When sensor is tested in 400 ppm water, it is observed that V1 and V2 increased after 10 mm ice formation as the signal is cut off. When the ice formation is 12 mm the V levels are highest, since the signal is cut off completely by ice (Table 8.4 and Fig. 8.9).

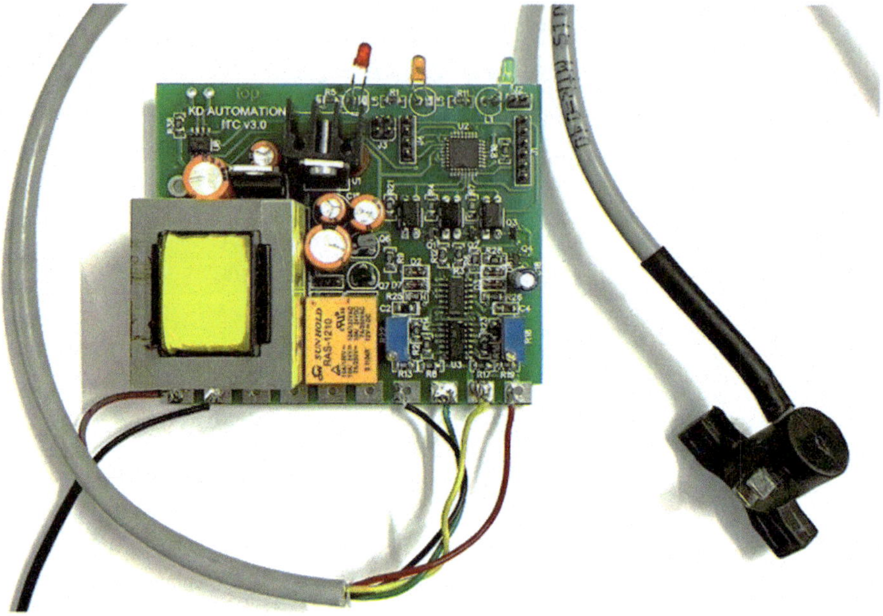

Fig. 8.7 Developed ice thickness sensor integrated with controller circuit

Table 8.3 Observations for TDS level 0–200 ppm

Time (min)	V1 (red and green)	V2 (yellow and black)
0	2.02	1.13
20	2.24	1.18
35 (10 mm)	4.56	1.26
60	6.33	1.315
80	6.9	1.75
100 (12 mm)	7. 29	4.8

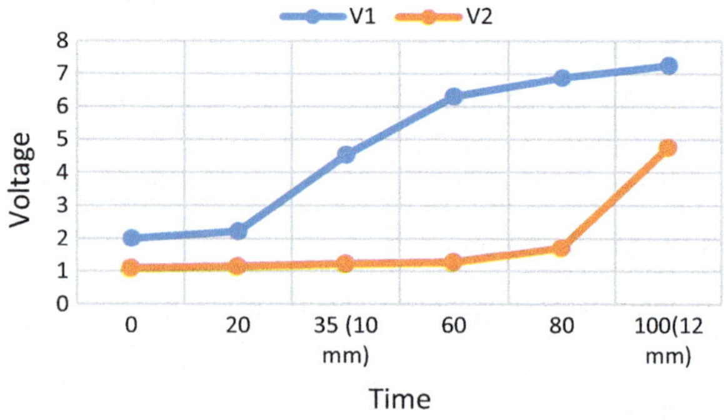

Fig. 8.8 Voltage output of sensor against time against time for TDS up to 200 ppm

Table 8.4 Observations for TDS level 400 ppm

Time (min)	V1 (red and green)	V2 (yellow and black)
0	2.07	1.63
20	2.09	1.33
40 (10 mm)	2.23	1.163
60	2.82	1.377
80	5.28	1.9
100 (12 mm)	6.59	3.97

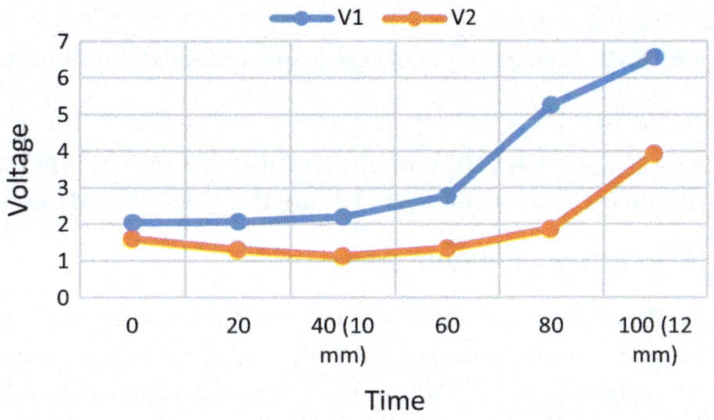

Fig. 8.9 Voltage output of sensor against time against time for TDS 400 ppm

TDS Level 600 ppm

When sensor is tested in 600 ppm water it is observed that V1 and V2 increase after 10 mm ice formation as the signal is cut off. When the ice formation is 12 mm the V levels are maximum (Table 8.5 and Fig. 8.10).

TDS Level 800 ppm

In this case, when sensor is tested for water having TDS level 800 ppm. The time taken for ice formation was lesser than previous case, since the TDS level of water is

Table 8.5 Observations for TDS level 600 ppm

Time (min)	V1 (red and green)	V2 (yellow and black)
0	2	1.235
20	2.11	1.247
40	2.11	1.308
60	2.37	1.347
80	2.84	1.399
100 (10 mm)	4.72	1.741
120	6.37	1.95
140	7.15	3.21
150 (12 mm)	7.41	5.03

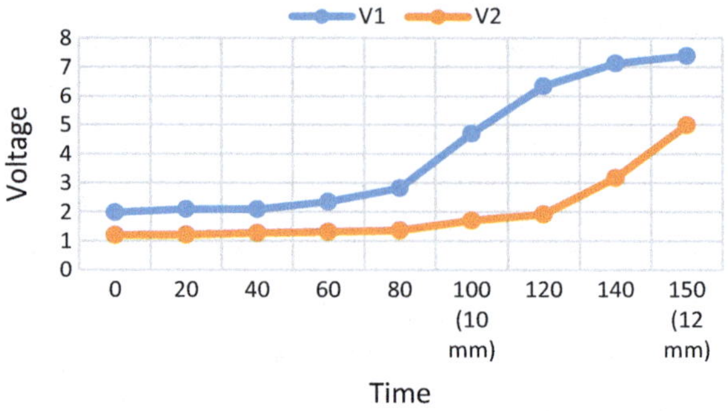

Fig. 8.10 Voltage output of sensor against time against time for TDS 600 ppm

increased. Here, we can observe that V1 and V2 increase after 10 mm ice formation. When the ice formation is 12 mm the V levels are maximum (Table 8.6 and Fig. 8.11).

TDS Level 1000 ppm

In this test, time taken for ice formation was observed to be greater than previous case since the ice formation level was zero in the beginning. We can see the similar trend in increase of V1 and V2 as in the previous test cases (Table 8.7 and Fig. 8.12).

Table 8.6 Observations for TDS level 800 ppm

Time (min)	V1 (red and green)	V2 (yellow and black)
0	2.31	1.34
20	2.71	1.52
40 (10 mm)	4.5	1.74
60	6.82	3.72
65 (12 mm)	6.9	4.1

Fig. 8.11 Voltage output of sensor against time against time for TDS 800 ppm

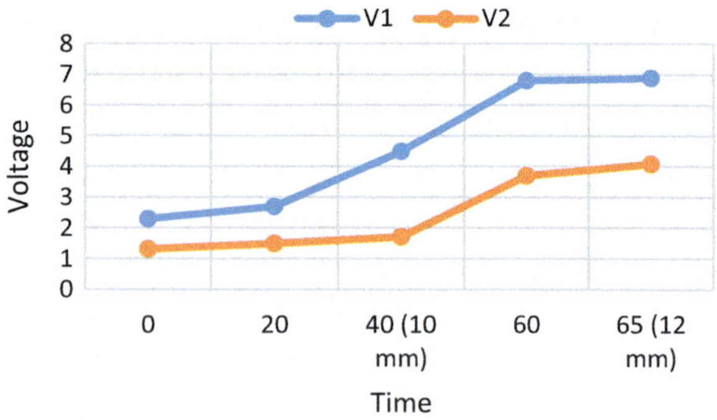

Table 8.7 Observations for TDS level 1000 ppm

Time (min)	V1 (red and green)	V2 (yellow and black)
0	1.91	0.653
20	2.06	0.706
40	2.07	0.648
60	2.12	0.673
80	2.21	0.729
100	2.54	0.928
120	2.92	1.267
140 (10 mm)	5.68	1.562
160	6.48	2.01
170 (12 mm)	7.02	4.47

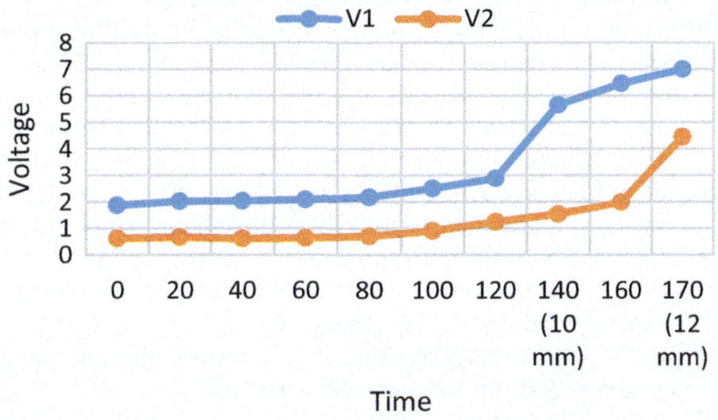

Fig. 8.12 Voltage output of sensor against time against time for TDS 1000 ppm

8.6 Conclusion

The Ice Thickness Controller has been successfully de-signed and implemented in Micro Can Chiller system. The goal was to make the Milk Can Chiller more efficient and energy saving with automated cooling. The new sensor was designed which overcame the disadvantages of previous system which included only two probes on the sensor which eventually ended up consuming more energy. The control circuit was also designed and implemented for the new sensor. Designed sensor is tested for different TDS level of water and gives consistent results for all the test cases.

The Ice Thickness Controller reduces the electricity usage by preventing the compressor to be ON for long time. The compressor in the system shuts OFF as soon as the required ice level is reached; this results in several on/off cycles, which may harm the compressor. Thereby, study of this impact and necessary actions can be undertaken in future work. Adding a time delay to the compressor's turning on and off after the level is reached can be implemented as a temporary remedial action.

References

Ajmal M, Nadeem M, Imran M (2018) Impact of immediate and delayed chilling of raw milk on chemical changes in lipid fraction of pasteurized milk. Lipids Health Dis 17:190–190

Bell T (2021, August 3) Electrical conductivity of metals ThoughtCo. https://www.thoughtco.com/electrical-conductivity-in-metals-2340117

Buddhi D, Ravi RK (2020) Thermal performance evaluation of a bulk milk cooler built-in cold thermal storage. J Crit Rev 7(15). ISSN: 2394-5125

Burke N, Zacharski KA, Southern M, Hogan P, Ryan MP, Adley CC (2018) The dairy industry: process, monitoring, standards, and quality. Descr Food Sci (Intechopen). https://doi.org/10.5772/intechopen.80398

Evans AC (2016) The bacteria of milk freshly drawn from the normal udders. J Infect Dis 18(5):437–476

Ghewade DV, Sapali DSN, Kajale DSR (2007) Dynamic modeling of bulk milk cooler. In: 5th International conference on HEFAT 2007 fluid mechanics and thermodynamics heat transfer

Luís RAJ, Silveira ABCV, Mário ERM, Fellipe DC-M, Oliveira D (2018) pp 510–517. https://doi.org/10.1590/1809-4430-Eng.Agric.v38n4p510-517/2018

Matheson R (2015) https://news.mit.edu/2015/promethean-power-india-milk-chillers-0908

Ministry of Fisheries, Animal Husbandry & Dairying (2022) Milk production in India—the journey of India's dairy sector. https://pib.gov.in/FeaturesDeatils.aspx?NoteId=151137&ModuleId%20=%202#:~:text=The%20per%20capita%20availability%20of,grams%20per%20day%20during%202021. Accessed 07 Sept 2022

Paludetti LF, Kelly AL, O'Brien B, Jordan K, Gleeson D (2018) The effect of different precooling rates and cold storage on milk microbiological quality and composition. J Dairy Sci 101(3):1921–1929. ISSN: 0022-0302. https://doi.org/10.3168/jds.2017-13668

Prakash R, Ravindra MR (2023) Cold thermal energy storage for milk chilling: a numerical and experimental study. J Food Eng 337

Thaker N, Tripathi N (2019) https://www.forbesindia.com/article/innovation-factories/promethean-power-a-chill-pill-for-dairy-farmers/53087/1

Thirstrup C, Deleebeeck L (2021) Review on electrolytic conductivity sensors. IEEE Trans Instrum Meas 70:1–22, Art no. 1008222. https://doi.org/10.1109/TIM.2021.3083562

Zakeri A, Saberi M, Hussain OK, Chang E (2018) An early detection system for proactive management of raw milk quality: an Australian case study. IEEE Access 6:64333–64349. https://doi.org/10.1109/ACCESS.2018.2877970

Chapter 9
Real-Time Monitoring and Battery Life Enhancement of Surveillance Drones

Pooja Kumari, Harshith Sourav Gosula, and Netra Lokhande

Abstract Surveillance drones have become indispensable instruments in a variety of businesses, including asset inspections, detection of breaches, and process monitoring. They are critical in both military and civilian applications, like search and rescue missions, border surveillance, animal monitoring, and even detecting problems in power systems such as the Corona effect. Despite substantial developments in drone technology, there are various obstacles in the drone architecture that limit its practical application. These difficulties include the inability to process data in real-time, which causes delays in decision-making, as well as issues such as low battery life, which restricts the amount of time a drone can stay in the air. Furthermore, drone propeller noise can be distracting, and scaling up drone operations for larger areas may be difficult due to coordination issues. This research paper addresses these issues by introducing creative approaches and novel ideas to improve surveillance drone architecture. It discusses cutting-edge technology such as edge AI, which involves establishing artificial intelligence directly on drones, innovative drone propeller designs, various AI algorithms, and fresh approaches to drone architecture. These improvements aim to make drones capable of performing difficult and vital jobs in addition to ordinary surveillance. The ultimate goal is to improve the efficiency and practicality of surveillance drones, allowing them to succeed in a variety of applications and circumstances.

Keywords Surveillance drone · Edge AI · Toroidal propeller · Battery life enhancement

P. Kumari · H. S. Gosula · N. Lokhande (✉)
School of Electrical Engineering, Dr. Vishwanath Karad MIT World Peace University, Pune 411038, India
e-mail: netra.lokhande@mitwpu.edu.in

P. Kumari
e-mail: pujji0210@gmail.com

H. S. Gosula
e-mail: suryahsourav@gmail.com

C. B. Khadse et al. (eds.), *Intelligent Methods in Electrical Power Systems*, Engineering Optimization: Methods and Applications, https://doi.org/10.1007/978-981-97-5718-3_9

9.1 Introduction

Efficiently keeping watch over vast areas and structures is a big challenge in many areas like security, event management, and public safety. It's also a concern in engineering, especially in power systems where high voltage transmission and distribution line maintenance and inspection may be a challenging, hazardous, and expensive undertaking. Drones drastically cut the costs of power line inspections for utilities by using their image processing capabilities. They also make things safer, more dependable, and quicker for fixing problems in the energy system, efficiently identifying threats to the energy grid, and dealing with things like the Corona effect in power systems. When electrical conductors are surrounded by ionized air, a phenomenon known as the corona effect occurs. This causes the conductors to glow and hiss. This not only reduces the efficiency of power lines but also damages the conductors. When power stations notice this effect, they have to temporarily shut down and make changes, like increasing the size of the conductors and the space between them. Drones are capable of identifying and reporting corona effects efficiently with their image processing capabilities.

The demand for surveillance extends beyond power systems. In India, for instance, there's a large workforce dedicated to keeping an eye on properties, buildings, and businesses. India has about 8.9 million private security guards and 1.9 million police officers. This means there are nearly five security guards for every police officer. The reason behind this is the growing need for private security and surveillance cameras. This need is increasing because people are moving to cities, there are higher crime rates, and there is a desire for better efficiency in factories and industries. This has forced a lot of security personnel to work overnight shifts, which are proven to be inefficient in many cases due to faults like human error, the need for breaks and rest by humans, delayed response times, and many more. For the same reason, drone surveillance is chosen as an effective alternative to human surveillance, as drones offer the ability to capture real-time data even from places that are difficult to access by humans. In fact, the drones have become so advanced that they could automatically avoid obstacles and chart their course, which means they could fly themselves without any human support when needed by using algorithms like Kalman filtering, which, when integrated with GPS and INS (Inertial Navigation System) data, enables the optimal estimation of aircraft position, velocity, and attitude by iteratively refining predictions based on the system's dynamics and sensor measurements, resulting in highly accurate and reliable navigation information (Werries and Dolan 2018).

Surveillance drones are one of many drones that are used for monitoring purposes. Technology and Artificial intelligence have greatly advanced, allowing them to identify criminal activity and fights between civilians by using algorithms like scatterNet Hybrid Deep Learning Networks (Singh et al. 2018). There are also times when drones are preferred over humans for monitoring purposes while dealing with large crowds, as seen in temples and events. But the use of surveillance drones for continuous monitoring of businesses and industrial operations is not feasible because of their flight time, which is usually around 8–13 min, and shorter battery size; they are meant to be small and fast (Sharma et al. 2020). So, despite the great advancements in drone technology and software algorithms, the practical usage of surveillance drones became restricted. This paper presents a novel idea and various methodologies which greatly improve the functioning efficiency and flight time of surveillance drones, making them feasible for continuous surveillance in places like power stations, industries, large environments, businesses, etc. Technologies like Edge AI, unique designs of propeller blades, and upgraded surveillance drone architectures are used to address the aforementioned problems.

Edge AI, in particular, represents a relatively recent development in artificial intelligence. It involves the deployment of AI models directly onto edge devices like microcontrollers and sensors located on or near the surveillance drone itself. For a better understanding, more about edge AI will be discussed in later sections. Section 9.2 of this paper discusses the standard architecture for surveillance drones. Section 9.3 includes the literature review and discusses current trends in drone technology. Section 9.4 will discuss the problems identified in the current drone architecture, and Sect. 9.5 will discuss proposed solutions. Section 9.6 will present the results and designs, and Sect. 9.7 will conclude the paper.

9.2 Surveillance Drone Architecture

Surveillance drones are advanced systems designed for real-time data collection and analysis. They offer a significant advantage by covering large areas much faster than ground-based personnel. These drones have a multi-layered architecture, combining hardware, software, and communication components to ensure efficient performance. They are equipped with various sensors, including cameras, thermal imaging devices, and sometimes LIDAR, or radar systems, to capture visual and environmental data. Onboard processing units, like microcontrollers and GPUs, handle preliminary data analysis and feature extraction. They are powerful tools with a well-integrated system for effective data collection and analysis in a variety of applications.

Currently, surveillance drones play an essential role. They act as intruder alert systems, swiftly alerting the security department when incidents like robberies or break-ins occur. Moreover, these drones function as scene analysis systems, predicting activities within their areas, including fights and crimes, using advanced technology. However, the practical application of such systems is not possible without incurring huge expenses. Though the basic parts for all drones are identical. The specifications of the components determine the use of the drone. Figure 9.1 shows a flow chart of the drone component's specifications that takes surveillance drones into account.

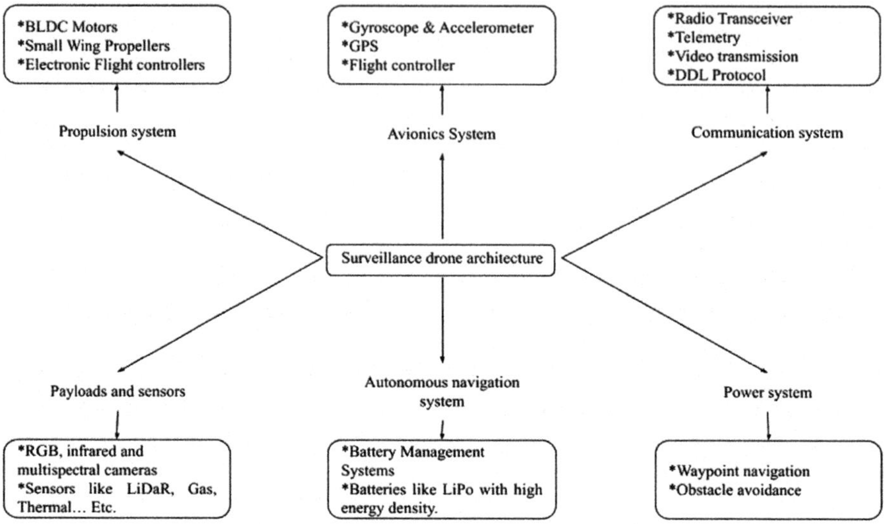

Fig. 9.1 Surveillance drone architecture

9.2.1 Propulsion System

The ability of a surveillance drone to fly is largely dependent on its propulsion system. In order to generate thrust, it frequently uses electric motors in conjunction with propellers, which enables the drone to maneuver, ascend, and descend smoothly. Many small UAVs commonly employ brushless DC motors or permanent magnet synchronous motors as their power sources due to their superior attributes, including exceptional efficiency, high energy density, strong reliability, rapid speed capabilities, and ease of control. The electric propulsion system in unmanned aerial vehicles (UAVs) powered by electric motors offers several benefits, including reduced carbon emissions, minimal pollution, cost-effectiveness, and efficient energy utilization. Electric propulsion systems in UAVs commonly make use of high-energy-density permanent magnet motors. Furthermore, when a high-power motor system is divided into several lower-power motor systems, each contributing to the same total power, the overall system maintains its power density and efficiency. This property, known as the relative scale-independent characteristic of motors, permits the employment of multiple lower-power motors to drive smaller fans with reduced diameters, leading to an enhanced thrust-to-weight ratio for the propulsion system. This, in turn, results in improved UAV stability and enables the implementation of an optimized energy management strategy for the UAV (Zhang et al. 2022).

9.2.2 Avionics System

Avionics is an intricate electronic system employed in aircraft, serving diverse functions including navigation assistance, communication establishment, and flight data recording, among others. This system encompasses various components, including flight controllers responsible for motor speed coordination, GPS modules for positioning the drone, gyroscopes that ensure balance and flight stability, accelerometers utilized for determining the drone's position and orientation during flight, and actuators employed for precise position adjustments. Ultimately, these systems are used to ensure stable flight, calibration, precise positioning, and efficient control of the drone's movement.

9.2.3 Communication System

These systems are responsible for the control of drones from the Ground Control Station (GCS) and also for the transfer of data from the drone. Surveillance Drones utilize various channels, like radio frequencies and Wi-Fi, while prioritizing encryption and reliability. The modern communication networks used in drones involve communication modules, IoT-enabled UAV communication systems, mobile edge computing, integrating UAVs with cellular networks, and UAV-assisted WSN and V2V (Ingle et al. 2022). Telemetry, as depicted in the above diagram, is a technique for automatically collecting, transmitting, and measuring data from remote sources using sensors.

9.2.4 Power System

The power system of surveillance drones includes batteries with varying capacities and voltages. The batteries are selected based on considerations like discharge rate, battery management systems, and charging time's impact on performance. Redundancy, energy efficiency, and safe battery handling are essential for extended flight times and reliable operation. Lithium-ion cells (LiPo) and Nickel-metal hydride cells (Ni-MH) are preferred as batteries because LiPo batteries have an excellent charge/discharge cycle capacity of 99.8% and Ni-MH has an excellent energy density of 360 MJ/m^3 (Adamski 2017).

The battery management system is another important part of the power system. Its main purpose is to keep the battery within the safety operation region in terms of voltage, current, and temperature during the charge, the discharge, and in certain cases, at open circuit. Rule-based, intelligence-based, and optimization-based energy management systems are generally used in drones based on the scenario, and the power system of a surveillance drone can hold components like fuel cells and supercapacitors other than batteries (Pham et al. 2022).

9.2.5 *Payloads and Sensors*

Payloads are the additional weights that the UAVs carry other than the flight systems. These also include cameras and a spectrum of sensors for various purposes. For example, Barometer for pressure, GPS sensor for coordinates, Distance sensors for obstacle avoidance, Magnetometer for identifying electronic and magnetic objects, etc.

9.2.6 *Autonomous Navigation*

Advanced drones are equipped with autonomous navigation systems such as way point navigation, where the drone touches certain way points or via points before reaching its destination. This is a path planning methodology. And they are also equipped with the "Return to Home" (RTH) methodology, where the drone returns to home on the same path it traveled in case of any failure or when it loses connectivity with the ground control station.

9.3 Current Trends of Drone Technology

The drone has a spectrum of sensors mounted on it. They are used for both adjusting the plane's flight and collecting important data. Sensors like an accelerometer and gyroscope are required in order to maintain a stable flight by calibrating its height, speed, and orientation. This part of adjusting the drone's flight by itself is called onboard processing because the drone handles this processing by itself without transmitting the data to the ground control station. The data from the other sensors is transmitted to the ground control station using technologies like WiFi for short range and telemetry, analog and digital radio, and cellular communications for operational ranges extending to hundreds of kilometers.

After the transmission is done, the data is locally stored and will be sent to post-processing, where different algorithms are used, like the Dijkstra algorithm for generating the shortest path for the drone, gradient descent optimization algorithms, and various sensing algorithms. After the processing is done in the ground control station, the insights from the data will be displayed to the operator in the form of graphs, maps, and image data, through which the operator can make a decision. This command will be transmitted back to the drone, and relevant actions will be

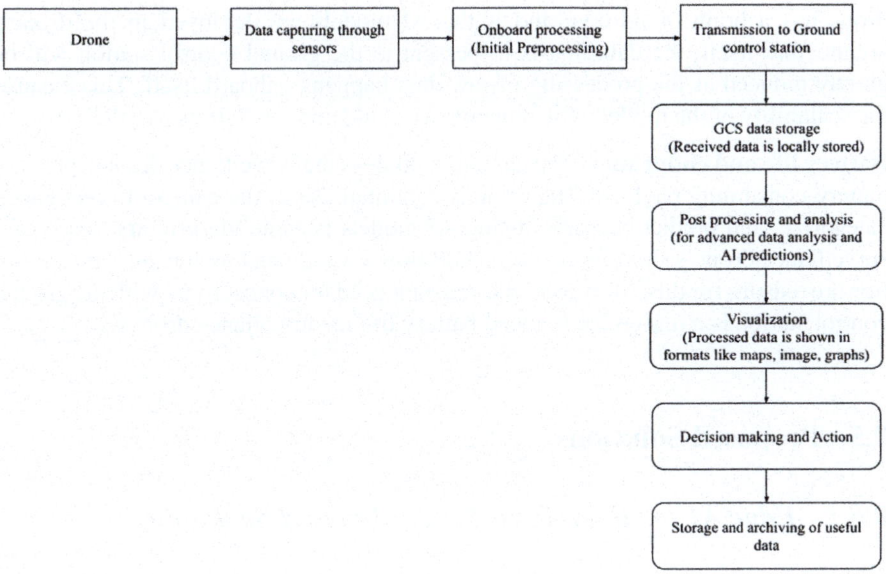

Fig. 9.2 Workflow of surveillance drones

performed. The insights are stored locally again for further usage and processing. All the aforementioned processes are described in Fig. 9.2.

9.4 Problems Identified with This Model

Real-time processing: Although there are systems like RTOS (real-time operating systems) that are being deployed in many applications like the Internet of Things (IoT), automotive and medical systems are improving efficiency and AI models that can classify, identify, and predict different kinds of scenarios, the drone technology is unable to take decisions in real-time, as we call it (Kangunde et al. 2021). The data needs to be transmitted to the ground control station, and the operator needs to understand and give relevant commands in order for the drone to perform the tasks. This makes the complexity of data processing very high. In addition to that, we cannot always coordinate with the ground control station because of the anti-drone technologies that will manipulate and block the drone's online connectivity, like radar blocking systems, GPS blocking systems, etc. Ultimately, drones are not functional unless there is an active connection with the control station.

Scalability: A single drone is usually not used for surveillance of an area; instead, a network of drones is used for monitoring and surveillance purposes So, if many drones are coordinating with a singular ground control station, it would add to the complexity of data management, processing, transmission, and speed. But if every drone has a brain of its own and if the AI models are deployed in the drone's architecture itself, the data that is to be sent to the ground control station will be greatly reduced as the processing of raw data happens onboard itself. This ensures the scalability of the model, which means we could add more drones to the network.

Battery life and endurance: The camera module is the surveillance drone's primary battery-consuming payload. The drone's continual use of the camera to recognize, categorize, and predict scenarios using AI models became the primary cause of a surveillance drone's short flying time. Additional weight and high battery consumption are reasons for this. Moreover, the constant need for connectivity with the ground control station is a reason for reduced battery life in surveillance drones.

9.5 Proposed Solutions

9.5.1 Edge AI for Real-Time Processing and Scalability

Introduction of AI into Drone Technology: In the early 2000s, cloud computing emerged as a novel computing infrastructure for the internet, with the main advantage of providing unlimited storage capacity and computing resources at a lower cost and carbon footprint. Cloud technology has paved the way for AI by facilitating storage and data transfer, as well as providing computing resources while the AI's algorithms for prediction and analysis are running. However, there were concerns about security and service speed because some processes required high-speed data processing, which was not possible with this architecture. This created a demand for a novel methodology in which no data is transferred via the cloud, paving the way for edge AI, which is a combination of AI and edge computing in which AI models are present in end devices such as microcontrollers and sensors, eliminating the need for data transfer. Edge AI is the practice of performing AI computations at the network's edge rather than in a central location such as a cloud service provider's data center or a company's own private data warehouse (Singh and Gill 2023). Because the AI models are present in the end devices, latency is reduced and security is increased. And by using the Edge AI architecture, the RAM footprint can be as low as 1 kB, making it the most optimized architecture available. This is achieved as no data is being transmitted to the consumer's network or device, as discussed above.

It improves bandwidth efficiency, which is critical in environments with limited connectivity, and it enables offline operation, which is critical in remote scenarios (Lee et al. 2018). It also saves energy, extends device battery life, enables adaptive decision-making in dynamic environments, and contributes to a decentralized AI ecosystem by distributing tasks across devices, increasing scalability. Furthermore, using this architecture, a group of drones can easily coordinate with one another without the need for a centralized network. All the aforementioned advantages of edge AI architecture are described in Fig. 9.3. Additionally, when surveillance drones are being used for high-security applications with normal AI, it can be susceptible to anti-drone systems like GPS jamming, Radar jamming, Radio frequency jamming, etc. When paired only with GPS for navigation, the enemy groups can also manipulate the incoming GPS or RF signals, changing the path and target of the drones to some other location. This method is known as drone spoofing (Park et al. 2021).

Training the Edge AI Model

The model training for this project has been done on Nano Edge AI Studio, which is a product of the STM microcontroller ecosystem. The microcontroller board, called the STM32 Nucleo F411RE, is chosen as the microcontroller.

Steps for training the model in Nano Edge AI Studio:

1. **Entering the Board Specifications**

The user must enter the information about the board they are using so that the model will be prepared accordingly. They also need to set the maximum flash usage and RAM utilization for that project, which is described in Fig. 9.4.

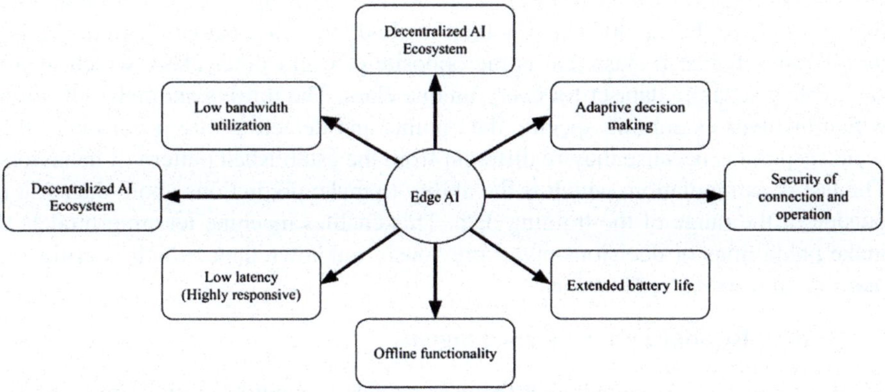

Fig. 9.3 Advantages of edge AI

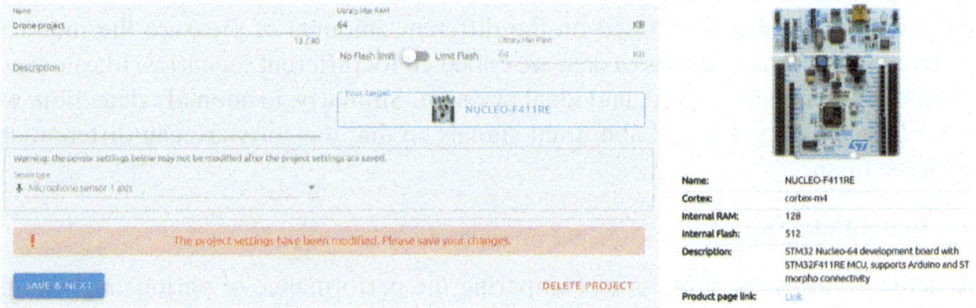

Fig. 9.4 Entering the board specifications

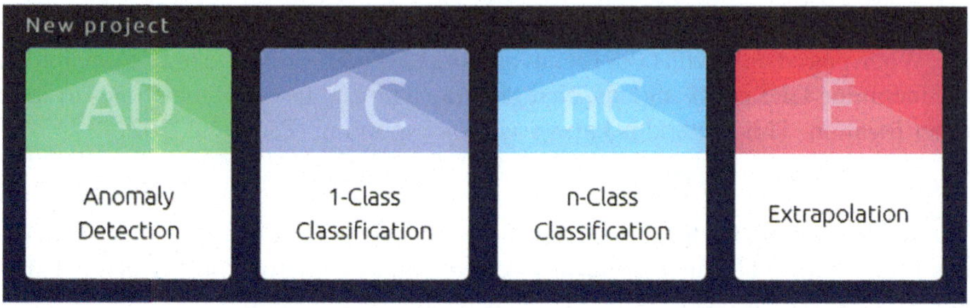

Fig. 9.5 Different approaches for model preparation in Nano Edge AI Studio

2. **Choosing a Model Preparation Approach**

There are four different kinds of approaches available in the software, as shown in Fig. 9.5. They are one-class classifications, which involve fitting a model to "normal" data and predicting whether new data is normal or an outlier. The second is N-class classifications, in which we must generate N*(N − 1)/2 binary classifier models for fitting N classes. Using this classification method, we can separate the main dataset into datasets for each class that is the opposite of every other class, which means we create a separate dataset for every unique class. The third is anomaly detection, which involves examining specific data points and detecting rare occurrences that seem suspicious because they're different from the established pattern of behaviors. The last is extrapolation, which is the ability to make predictions about data that is outside of the range of the training data. This enables machine learning models to make predictions or decisions about previously unknown data, which is critical in many deep learning applications.

3. **Feeding Regular and Abnormal Signals**

All the unique sets of signals that the user intends to identify or avoid must be fed to the model by creating different classes. For example, in the N-class classification approach, the user needs to feed the signals of all the different kinds of signals that they want by creating unique classes. Figure 9.6 demonstrates the vibrational behavior data of a motor based on the different amounts of pressure the motor is generating. Different classes of data are collected for different scenarios, like no load, overpressure, underpressure, and ideal pressure. Similarly, in anomaly detection, we must feed both regular and abnormal signals so that the software can differentiate between them.

4. **Benchmarking**

Benchmarking is the process of comparing the performance of various algorithms in distinguishing between normal and abnormal signals and correctly classifying the elements. In Nano Edge AI Studio, the benchmarking process automatically finds and optimizes the best algorithmic combination among more than 500 million possible

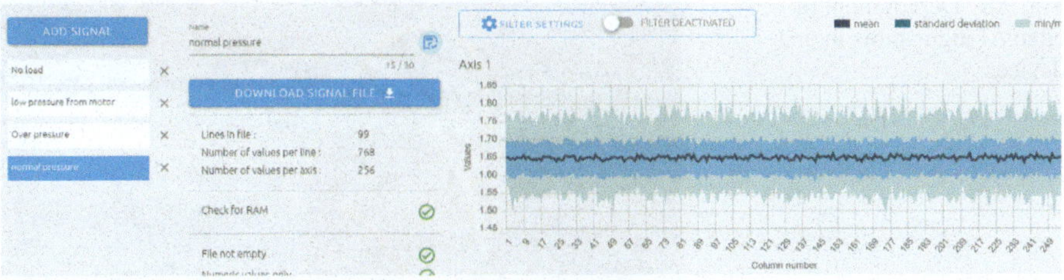

Fig. 9.6 Vibrational data of a motor at different pressures

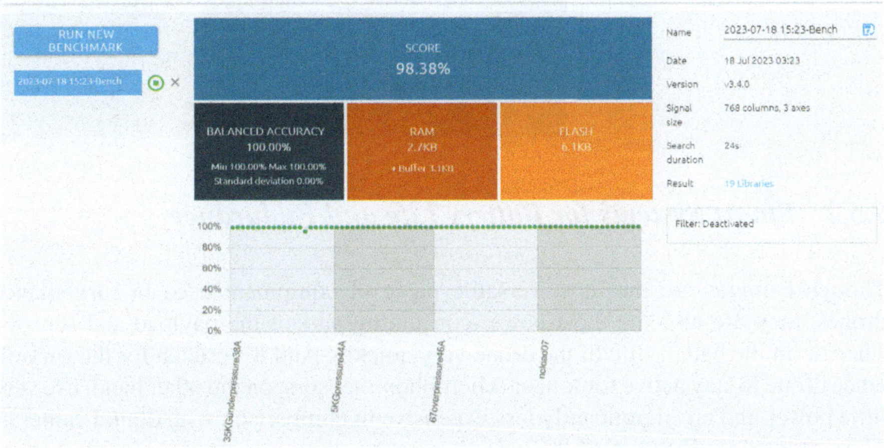

Fig. 9.7 Benchmarking process in Nano Edge AI Studio

libraries. The user can validate that the selected algorithm is the best fit for the given data by looking at the benchmarking score, which is described in Fig. 9.7.

5. **Deployment**

After selecting the algorithm using the benchmarking process, a library can be created directly from the software. This library can be used in various IDEs to code and deploy the algorithm into edge devices like microcontrollers and sensors. Figure 9.8 describes how the created library is being used in the Arduino IDE for practical deployment. Figure 9.8 shows how a microphone sensor, along with edge AI algorithms, is able to detect and classify sounds without any data transfer and minimal battery usage.

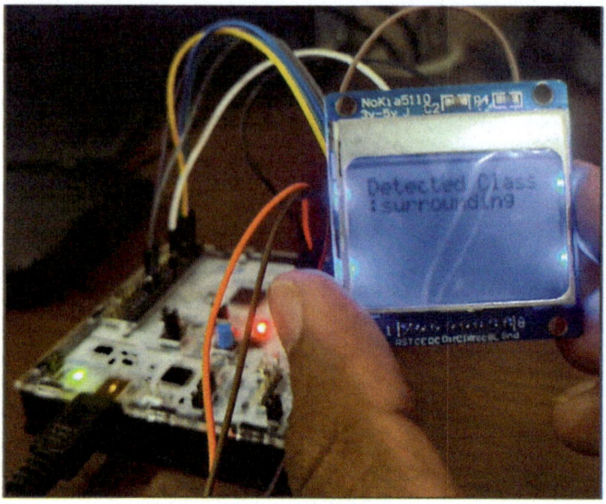

Fig. 9.8 Deployment of edge AI algorithms in end devices

9.5.2 Smart Systems for Battery Life and Endurance

Though cameras are the most versatile piece of equipment used in surveillance drones, they are also the most power-consuming part of the payload and sensors. They drain the battery life of the drone very quickly. And it's crucial for the surveillance drone to stay active for longer. Microphone sensors, on the other hand, use very little power and are significantly less expensive in comparison with digital cameras. Thus, strategic utilization of the camera must be done in order to save the battery and yet maintain performance by combining the low power-consuming ability of the microphone sensor with better AI model prediction accuracy to solve the problem of battery life and endurance. In other words, in order to successfully monitor an area in real-time, we must employ the capabilities of both the microphone and the camera.

Description of the Novel Approach

If we consider high security—large areas, as shown in Fig. 9.9, representing different blocks of a locality that also houses a shopping mall and a power station, we cannot rely solely on security guards to successfully monitor it because humans are prone to errors such as being careless, forgetting protocols, restlessness, and so on. We cannot rely solely on standard cameras because they are fixed in one position and lack AI models capable of detecting and reporting accidents. They are also inoperable in low-light conditions. And providing thermal imaging capabilities that work in low light conditions for all fixed position cameras is not cost effective. This is the kind of scenario that is tailor made for drones, as they can fly, be fast, be equipped with thermal cameras, and be more functional and efficient through the utilization of various algorithms.

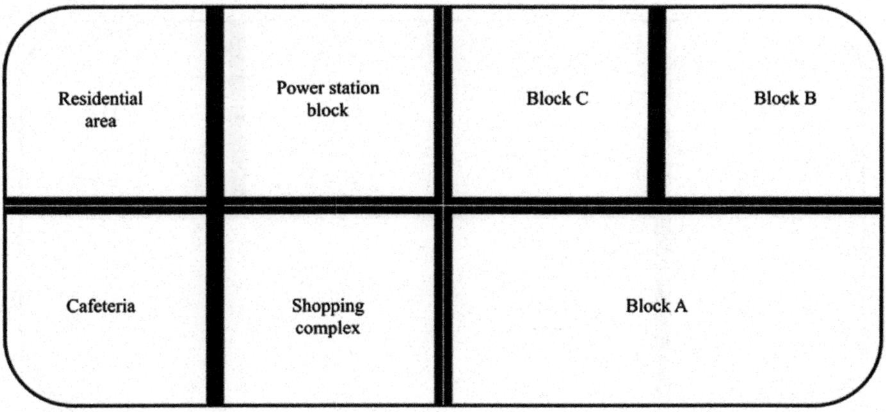

Fig. 9.9 Demonstration of a high security area

Steps

1. Using anomaly detection, we develop an edge AI model for microphone sensors. This model will help us detect sudden disturbances, such as someone walking or talking when the area is previously silent. It does not directly detect sounds but rather calculates the deviation of audio signals from what they were previously, which means that the model will be triggered by sudden disturbances in the sounds, such as a gunshot while everyone was doing their normal work. Sounds while someone is walking or talking, while it was dead silent just a few minutes ago. This will help the model detect scenarios that are not meant to happen.
2. We deploy the aforementioned model on the edge devices and place them all over the place such that the entire area will be covered with the range of the microphone sensors, as shown in Fig. 9.10, while the dotted blue circles represent the detecting range of microphone signals.
3. A series of drones will be stationed at key points throughout the area and will remain in standby mode until any disturbance or anomaly is detected by the microphone sensors. All the drones must be coded in such a way that when an anomaly is detected in a specific microphone's range, it should be able to fly there automatically without the use of camera. One such way to do so will be discussed in the next sub-section named "Method for Autonomous Flying."
4. When an anomaly is detected within a specific microphone's range, the nearest drone to that point will be deployed to that location, and upon reaching that location, the camera will be turned on, and edge AI algorithms will be detecting the scenarios in real-time and reporting them via local IoT devices to the security personnel when required. This aforementioned process is described in Fig. 9.11, and the block diagram for the microphone sensors is described in Fig. 9.12.

The anomaly is represented by a red dotted circle in Fig. 9.11. When an anomaly is detected within the range of a particular sensor, the neighboring sensor's sensitivity

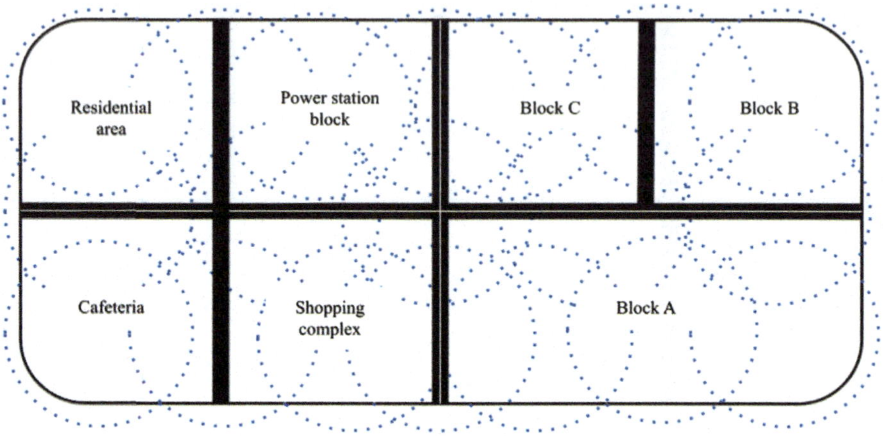

Fig. 9.10 Detecting range of microphone sensors

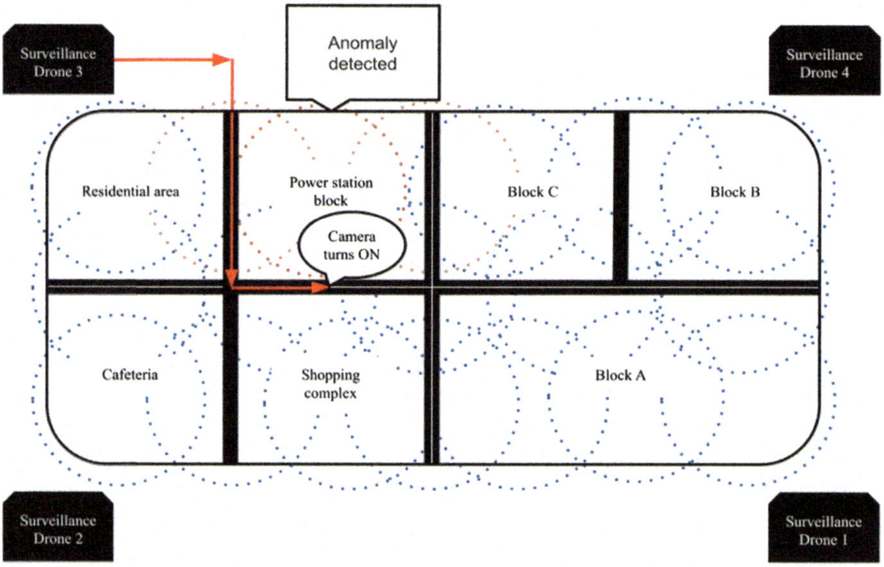

Fig. 9.11 Drone surveillance via a novel approach

will be increased for better results, and the other drones in standby will be deployed when needed.

Method for Autonomous Flying

Ideally, in order to code a drone, you have to connect to the drone via an IDE, set the telemetry and parameter information, and create a code with a standard set of commands in the sequence. For example, a simple code should have the commands

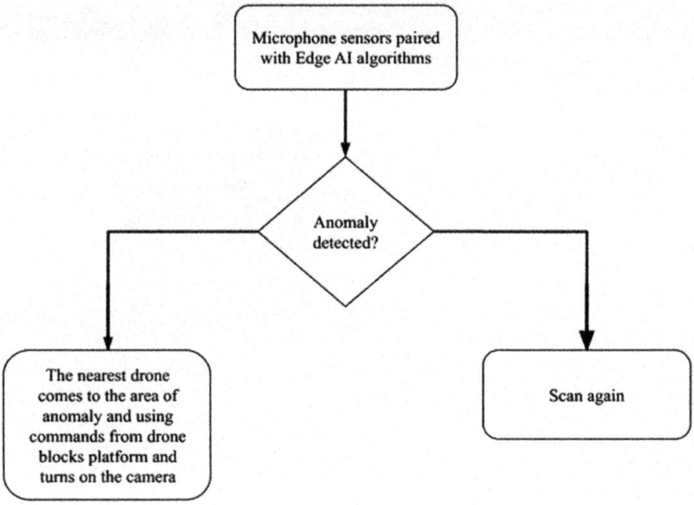

Fig. 9.12 Block diagram of the microphone sensor

that will start all the motors, go up by 10 m, move forward by 10 m, and turn on the camera. All the drones have to be coded in this way so that they can reach the area of the microphone where the anomaly has occurred without the help of the camera, ultimately increasing the battery life and endurance.

This can be done via multiple software such as DroneKit-Python, PX4, ArduPilot, DroneBlocks, etc., which is rather a simpler one in comparison with others. Figure 9.13 displays a screenshot of the DroneBlocks platform giving commands to the drone to reach the anomaly point, which is described in Fig. 9.11.

By following the aforementioned commands, the drone will go to the point of anomaly, and after its job is done, it will fly back to its landing pad as well.

There is just one drawback with this novel approach, which is that the sound of the drones might trigger the microphone sensor, giving a false result and it might also alert the intruders who came in. This problem can be solved in two steps, one is minimizing drone noise by using toroidal propellers, and the other is the MFCC scheme for neglecting the sound from drones by the microphone sensors.

Step 1: Toroidal propellers for drone noise reduction

The heart of noise reduction and low-height operations lies in the propeller design. By reducing air contact while maintaining consistent RPM, we can minimize the amount of sound emitted. This propeller design was made by the Lincoln Laboratory of the Massachusetts Institute of Technology, and it's named the toroidal propeller for its shape (Sebastian and Strem 2019). The toroidal propeller consists of two to three blades looping together so that the tip of one blade curves back into the other. This closed-form structure minimizes the drag effects of swirling air tunnels (i.e., vortices) created at the tips of blades and strengthens the overall stiffness of the propeller.

Features:

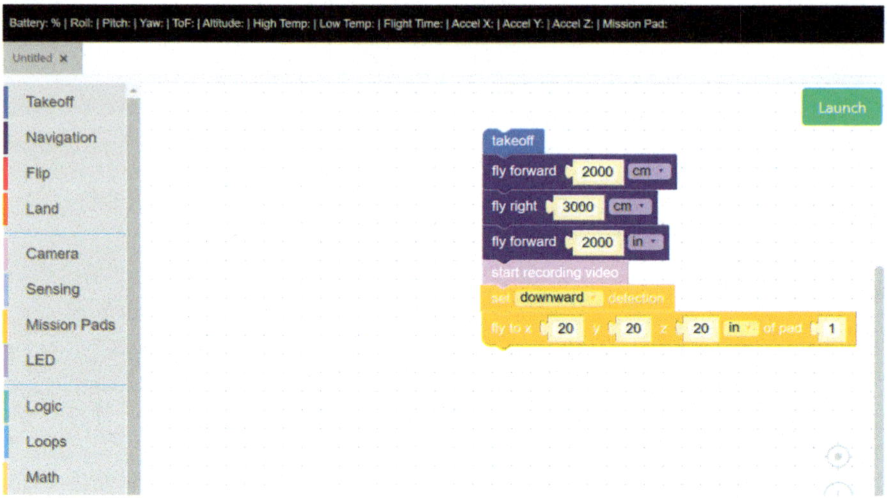

Fig. 9.13 Screenshot of commands given to the surveillance drone

- Reduces noise without requiring supplementary components that add weight and increase power draw.
- Reduces the likelihood that the spinning propeller will cut or clip objects or surfaces in the drone's path.
- Achieves thrust comparable to that of a multirotor drone propeller.

Step 2: MPCC Scheme

The MFCC (Mel Frequency Cepstral Coefficients) technique is a complex but effective approach to identifying, classifying, and neglecting sounds produced by drones in real-world, noisy conditions. MFCC is a widely accepted technique in audio processing that mimics how the human auditory system responds to sound (Shi et al. 2018). It involves several steps:

- Pre-emphasis: Enhancing high-frequency components in the audio.
- Framing: Breaking the audio signal into short time segments.
- Windowing: Applying a specific function to each time segment.
- Fast Fourier Transform (FFT): Transforming each segment into the frequency domain.
- Mel Filterbank: Employing triangular filters in the power spectrum to simulate human hearing.
- Logarithm: Calculating the logarithm of filterbank energies.
- Discrete Cosine Transform (DCT): Applying DCT to derive MFCCs, which represent the spectral characteristics of each segment.

By following the aforementioned steps while developing the algorithm, we can make the microphone sensor identify and ignore the sounds of the drones, preventing false triggering caused by drone noise as it passes by the microphone sensors during inspections.

9.6 Results

Deployment of the Edge AI Microphone Sensor Algorithm Using IDEs

The library created for the microphone sensor data on anomaly detection is deployed via Arduino IDE into the STM 32 Nucleo F411RE microcontroller, and the results simulate normal and abnormal conditions. The graphs of the data for both the abnormal and normal signals are displayed in Fig. 9.14. The data displayed in the graphs are collected manually and preprocessed before training the model.

When the normal surrounding noise was played near the microcontroller, the system detected it as normal signals, and when the sound was suddenly changed to gun shots, the system detected it as abnormal signals with almost 100% accuracy, as shown in Fig. 9.15a, b, respectively.

Apart from anomaly detection, the Edge AI algorithm was also deployed using the Arduino IDE and tested to classify different sounds like baby crying, surrounding noises, ambulances, gun shots, etc., which it successfully did. The results for the same are displayed in Fig. 9.16. Though the results were being displayed on the computer's serial monitor for display purposes, all the required algorithms and computations were run on the microcontroller itself.

Toroidal Propeller Versus Normal Propeller Decibel Reading

Figure 9.17 shows two standard propellers mounted on a DJI Tello drone in black and two toroidal propellers mounted on the drone in white for comparison. While

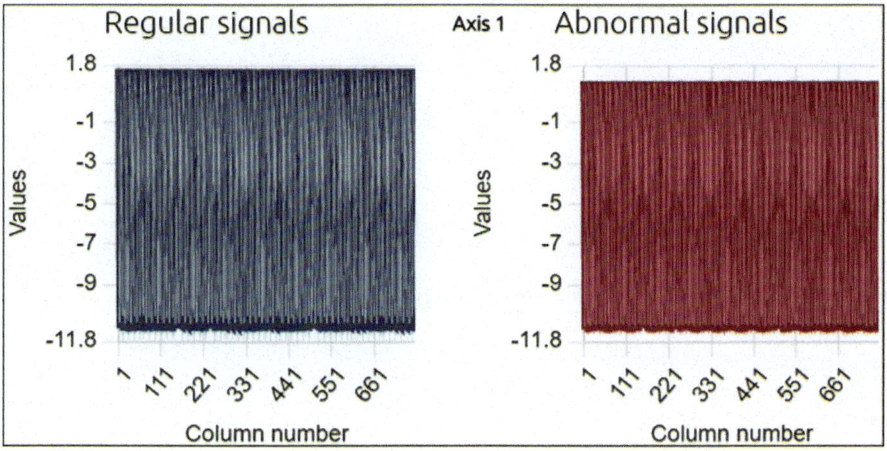

Fig. 9.14 Graphs of normal signal data and abnormal signal data

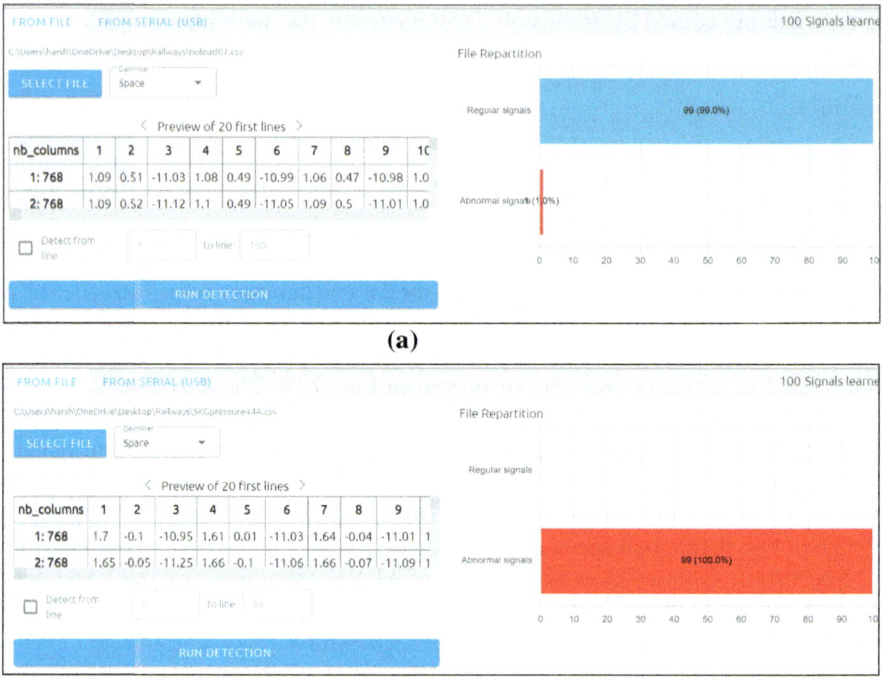

Fig. 9.15 **a** Output of model when normal signals are fed to the microphone. **b** Output of model when abnormal signals are fed to the microphone

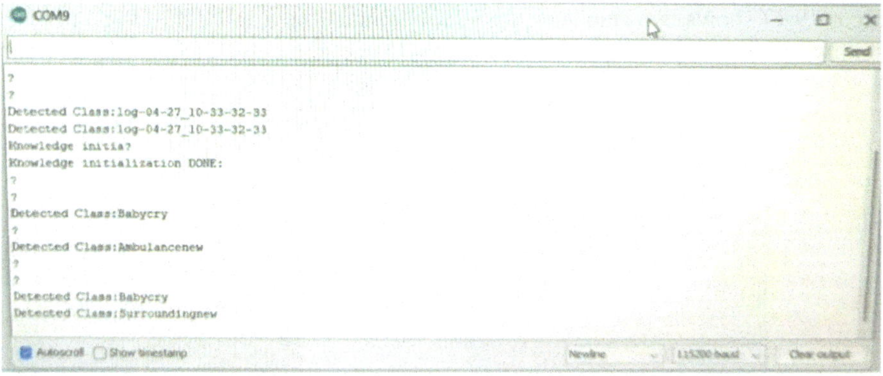

Fig. 9.16 Audio classification using edge AI algorithms

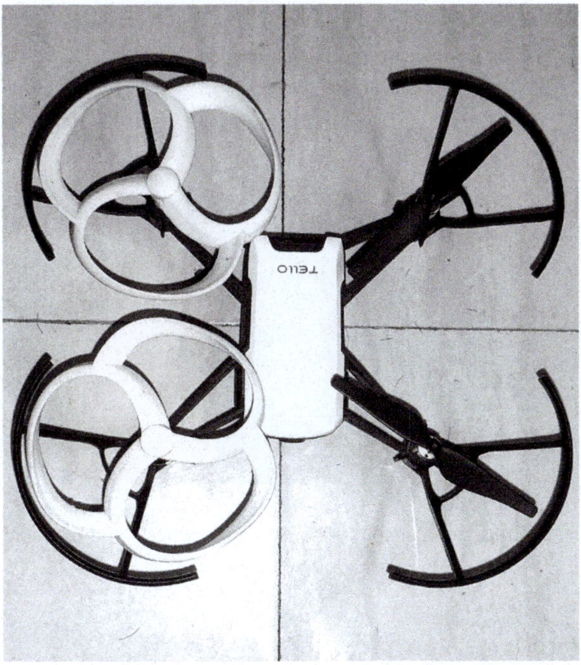

Fig. 9.17 Toroidal propellers and normal propellers mounted on Tello drone

the drone is operating at a constant 2000 rpm, we have taken the decibel readings from both propeller designs in comparison with surrounding noise and displayed the results in Fig. 9.18.

As shown in Fig. 9.18, the noise produced by the toroidal propellers is only 71 dB, compared to 101.1 dB by the normal propellers and 61 dB by the surrounding noise.

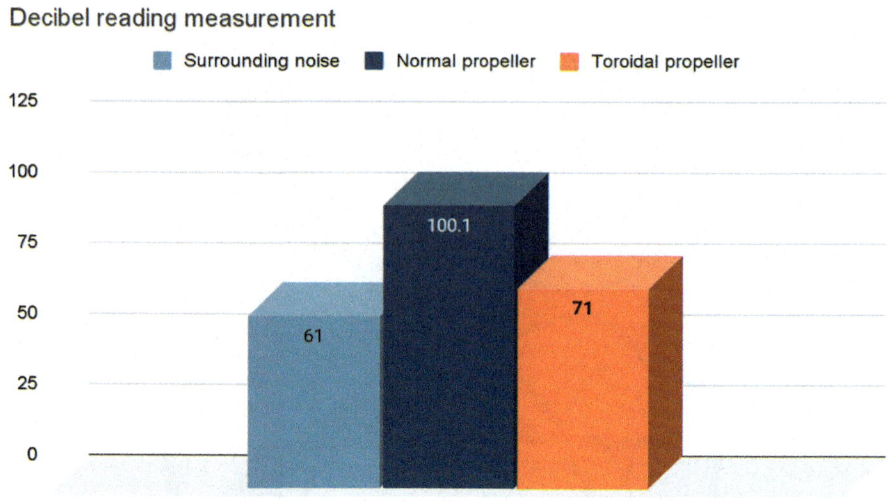

Fig. 9.18 Decibel reading comparison of normal propellers and toroidal propellers with surrounding noise

9.7 Conclusion

Surveillance drones offer a great advantage over monitoring large spaces and scenarios, but their practical usage was restricted because of problems in the drone architecture like low battery life, high noise from drones, inefficient communication protocols, and being unable to take actions in real-time.

There needs to be a shift in the drone architecture in order to make their practical usage feasible. Different technologies discussed in the above research must be included in the drone architecture to do so. The introduction of the Edge AI model will make surveillance drones take real-time decisions, function without any kind of data transfer, reduce energy usage, and also make them fast and secure for high-level applications, as discussed above.

Propeller designs like toroidal propellers will enable the drone to make low to minimal noise and enable them to operate in stealth mode because of their unique and innovative structure. To additionally save battery life while monitoring large spaces, the novel approach of strategic utilization of the camera must be taken in order to save the battery and yet maintain performance by combining the low power consuming ability of the microphone sensor with better AI model prediction accuracy to solve the problem of battery life and endurance by combining the capabilities of both digital camera sensors and analog microphone sensors. By making the aforementioned changes in the architecture, surveillance drone technology is made feasible for real-time practical utilization owing to better performance and results.

References

Adamski M (2017) Analysis of propulsion systems of unmanned aerial vehicles. J Mar Eng Technol 1–7. https://doi.org/10.1080/20464177.2017.1383337

Ingle PY et al (2022) DVS: a drone video synopsis towards storing and analyzing drone surveillance data in smart cities. Systems 10(5):170. https://doi.org/10.3390/systems10050170

Kangunde V, Jamisola Jr RS, Theophilus EK (2021) A review on drones controlled in real-time. Int J Dyn Control 9(4):1832–1846. https://doi.org/10.1007/s40435-020-00737-5. Epub 2021 Jan 5. PMID: 33425650; PMCID: PMC7785038

Lee Y-L, Tsung P-K, Wu M (2018) Technology trend of edge AI. In: 2018 International symposium on VLSI design automation and test (VLSI-DAT), Hsinchu, Taiwan, pp 1–2. https://doi.org/10.1109/VLSI-DAT.2018.8373244

Park S, Kim H, Lee S, Joo H, Kima H (2021) Survey on anti-drone systems: components, designs, and challenges. IEEE Access 1. https://doi.org/10.1109/ACCESS.2021.3065926

Pham KL, Leuchter J, Bystricky R, Andrle M, Pham NN, Pham VT (2022) The study of electrical energy power supply system for UAVs based on the energy storage technology. Aerospace 9:500. https://doi.org/10.3390/aerospace9090500

Sebastian T, Strem C (2019) Toroidal propeller, US 2019/0135410 A1

Sharma A, Vanjani P, Paliwal N, Wijerathna B, Chathuranga M, Jayakody DN, Wang H-C, Muthuchidambaranathane P (2020) Communication and networking technologies for UAVs: a survey. https://doi.org/10.36227/techrxiv.12912449.v1

Shi L, Ahmad I, He YuJing, Chang K (2018) Hidden Markov model based drone sound recognition using MFCC technique in practical noisy environments. J Commun Netw 20:509–518. https://doi.org/10.1109/JCN.2018.000075

Singh R, Gill SS (2023) Edge AI: a survey, internet of things and cyber-physical systems, vol 3, pp 71–92. ISSN: 2667-3452. https://doi.org/10.1016/j.iotcps.2023.02.004

Singh A, Patil D, Omkar SN (2018) Eye in the sky: real-time drone surveillance system (DSS) for violent individual identification using scatterNet hybrid deep learning network. In: 2018 IEEE/CVF conference on computer vision and pattern recognition workshops (CVPRW), Salt Lake City, UT, USA, pp 1710–17108. https://doi.org/10.1109/CVPRW.2018.00214

Werries A, Dolan JM (2018) Adaptive Kalman filtering methods for low-cost GPS/INS localization for autonomous vehicles. Carnegie Mellon University. Journal contribution. https://doi.org/10.1184/R1/6551687.v1

Zhang B, Song Z, Zhao F, Liu C (2022) Overview of propulsion systems for unmanned aerial vehicles. Energies 15:455. https://doi.org/10.3390/en15020455

The manufacturer's authorised representative in the EU is Springer Nature Customer Service Centre GmbH, Europaplatz 3, 69115 Heidelberg, Germany. If you have any concerns regarding our products, please contact ProductSafety@springernature.com

Printed and bound by CPI Group (UK) Ltd, Croydon, CR0 4YY
01/10/2026
02225919-0013